Making Your CAM Journey Easier with Fusion 360

Learn the basics of turning, milling, laser cutting, and 3D printing

Fabrizio Cimò

BIRMINGHAM—MUMBAI

Making Your CAM Journey Easier with Fusion 360

Copyright © 2023 Packt Publishing

Group Product Manager: Rohit Rajkumar

Publishing Product Manager: Vaideeshwari Muralikrishnan

Senior Editor: Hayden Edwards

Technical Editor: Joseph Aloocaran

Copy Editor: Safis Editing

Project Coordinator: Sonam Pandey

Proofreader: Safis Editing

Indexer: Tejal Daruwale Soni

Production Designer: Jyoti Chauhan

Marketing Coordinator: Nivedita Pandey

First published: March 2023

Production reference: 1170223

Published by Packt Publishing Ltd.

Livery Place

35 Livery Street

Birmingham

B3 2PB, UK.

ISBN 978-1-80461-257-6

www.packtpub.com

To my grandfather, Gaetano, 99 years old and counting; a living miracle of cleverness and goodness, and a constant source of inspiration throughout my life.

To my girlfriend, Giulia, who was the first to convince me to go back to studying engineering and achieve what remained just a dream for a long time.

To my parents, Francesca and Antonio, for having supported me on such a long and winding journey.

– Fabrizio

Contributors

About the author

Fabrizio Cimò is an Italian engineer who has always had a passion for the world of 3D graphics and design.

During his first degree in industrial design, he started uploading video lessons covering 3D modeling tools such as Rhinoceros, Blender, and Fusion 360 on his YouTube channel.

Sometime later, Autodesk noticed his work and asked him to keep promoting Fusion 360 as an official student ambassador.

During this time, he also joined Dynamis PRC, a racing team from Milano PT competing in the Formula SAE championship; thanks to this experience and his avid curiosity, he improved his knowledge of the manufacturing world.

Today, Fabrizio works as a machine designer for an important company in the laser-cutting sector.

I want to thank all my colleagues from the technical office for teaching me so much, with so much patience and passion, over all these years.

About the reviewers

J Aatish Rao is a dynamic, team-spirited mechanical engineering professional with more than 10 years of industrial experience. Certified to provide training on various software, science, and engineering-related topics, he is an Autodesk Certified Professional, Autodesk Fusion Certified User, ClearFit Certified Team Leader, and industrial consultant. Conducting workshops and engaging in teaching activities are the key ingredients of his life. He is also a keen enthusiast of stargazing.

Prashant Kumar is a graduate mechanical engineer with more than 7 years of industrial work experience. He specializes in engineering design software such as AutoCAD, SOLIDWORKS, and Fusion 360, with a good grasp of engineering subjects and principles.

Prashant currently works as a full-time freelance CAD designer, and he created a brand named 3Diest that provides engineering design and drafting services to companies and individuals across the globe. He also works as a part-time trainer for software such as AutoCAD and Fusion 360.

Nikhil Chaitanya is a mechanical engineer who completed his M.Tech in Machine Design specialization from Andhra University, Visakhapatnam, India. He has 3 years of industrial experience, with most of that time spent working at Tata Consultancy Services. He is currently working as an assistant professor in the Department of Mechanical Engineering at Vignan's Institute of Information Technology, Visakhapatnam.

Nikhil is passionate about new product development and building the most important manufacturing infrastructure for India with his start-up venture branded as the Namoona Group. He was also nominated for the Times of Startup India's 40 Under 40 Awards 2023 but has declined the award as he wants to encourage fellow students to become leaders of the future.

Table of Contents

Preface   xv

Part 1 – Implementing Turning Operations in Fusion 360   1

1

Getting Started with Turning and Its Tools   3

Technical requirements	4	Exploring main machining strategies	15
Approaching a lathe and its components	4	Longitudinal operations	15
Cylindrical coordinate system	5	Facing operations	16
Different types of turning operations	6	Plunging operations	16
		Profiling operations	17
Understanding the main parameters	8	Understanding tool geometry	18
Turning speed	9	Tool edges	18
Cutting speed	10	Tool surfaces	20
Cutting depth	11	Tool angles	21
Cutting feed	11	Summary	29
Cutting power	13		

2

Handling Part Setup for Turning   31

Technical requirements	31	Finding example projects in Fusion 360	33
Exploring the main interface of a CAM project	32	Importing our chuck	35

Understanding how to set up the first part to be machined   **39**

Setup tab   40

Stock tab   54

The Post Process tab   59

Summary   **61**

3

Discovering the Tool Library and Custom Tools   63

Technical requirements   64

Discovering the tool library   64

Creating a new tool   65

Getting a sample tool   66

Creating a new tool   67

Importing a third-party tool library   74

Importing new tools with a plugin (CoroPlus)   77

Summary   83

4

Implementing Our First Turning Operation   85

Technical requirements   85

Setting up a facing operation   86

Using CoroPlus to calculate cutting parameters   87

Entering the cutting parameters into Fusion 360   91

Discovering tool simulation   102

Checking the generated G-code   111

Settings   112

Operations   114

Summary   116

5

Discovering More Turning Strategies   117

Technical requirements   117

Turning Profile (Roughing)   118

The Tool tab   120

The Geometry tab   123

The Passes tab   124

Simulation   125

Turning Profile (Finishing)   126

The Tool tab   127

The Geometry tab   128

The Passes tab   128

Simulation   130

Turning Threads   130

The Tool tab   133

The Geometry tab   134

The Radii tab   136

The Passes tab   136

Simulation	138	The Passes tab	148
Drilling	**140**	Simulation	150
The Tool tab	140	**Turning Part**	**150**
The Geometry tab	141	The Tool tab	151
The Cycle tab	142	The Geometry tab	152
		The Radii tab	154
Turning Groove	**144**	The Passes tab	154
The Tool tab	145	Simulation	156
The Geometry tab	147	**Summary**	**156**

Part 2 – Milling with Fusion 360   157

6

Getting Started with Milling and Its Tools   159

Technical requirements	159	Cutting power and torque	168
Understanding what milling is and how it works	**160**	**Introducing the most common milling operations**	**170**
Cartesian machines	160	Face milling	170
Multi-axis machines	161	Shoulder milling	172
		Slot milling	173
Understanding the main cutting parameters	**162**	Profile milling	175
		Other	175
Spindle speed and cutting speed	162	**Summary**	**176**
Cutting depths	163		
Feed step	164		

7

Optimizing the Shape of Milled Parts to Avoid Design Flaws   177

Technical requirements	177	Backside milling	184
Handling undercuts and accessibility	**178**	**Learning how to manage mill radius**	**185**
Changing part orientation relative to the tool	180	Reducing the mill radius	186
Using a multi-axis machine	181	Tweaking our part geometry	188
Creating custom tools	182	Changing the cutting direction	189

Solving a very bad design	191	Counterbore holes	195
Undercut face	191	All around radii	197
Missing tool radii	194	Summary	197

8

Part Handling and Part Setup for Milling   199

Technical requirements	199	Stock tab	208
Understanding the part	200	Post Process tab	209
Choosing part placements	201	Defining the second setup	210
Choosing a part fixture	202	Setup tab	210
Choosing WCS offsets	204	Stock tab	211
Defining the first setup	206	Post Process	212
Setup tab	207	Summary	213

9

Implementing Our First Milling Operations   215

Technical requirements	216	Drilling	238
Face milling	216	The Tool tab	239
Using CoroPlus to find the best tool for face milling and shoulder milling	216	The Geometry tab	240
		The Heights tab	240
Implementing face milling with Fusion 360	224	The Cycle tab	242
Shoulder milling	229	Tapping	244
Calculating the cutting parameters by hand	229	Summary	247
Shoulder milling inside Fusion 360	233		

10

Machining the Second Placement   249

| Technical requirements | 250 | Implementing a roughing operation using adaptive clearing | 251 |
| Face milling | 250 | The Tool tab | 255 |

The Geometry tab   256
The Heights tab   256
The Passes tab   258
The Linking tab   260

Milling a hole   **262**
The Tool tab   263
The Geometry tab   264
The Heights panel   264
The Passes tab   265

**Finishing the part using a morphed
spiral**   **269**
The Tool tab   270
The Geometry tab   271
The Heights panel   272
The Passes tab   273

Thread milling   **275**
Thread geometry   276
Picking the proper threading tool   279
Implementing thread milling in Fusion 360   280

Summary   **286**

Part 3 – Laser Cutting Using Fusion 360   287

11

Getting Started with Laser Cutting   289

Technical requirements   **289**
Introducing lasers   **289**
How does a laser cut?   **291**

**Reviewing the pros and cons of laser
cutting**   **293**
Advantages of laser cutting   293
Drawbacks of laser cutting   294

Summary   **296**

12

Nesting Parts for Laser Cutting   297

Technical requirements   **297**
Presenting the example model   **298**
Understanding nesting optimization **299**
Sheet format   299
Batch volume   300

Creating a nesting with Fusion 360 **302**

Manual placement   302
The Arrange command   302
The Nesting and Fabrication extension 307

Summary   **319**

13

Creating Our First Laser Cutting Operation   321

Technical requirements	321	The Cutting data tab	329
Using Fusion 360 for laser cutting	321	**Implementing our first cutting operation**	**330**
Creating a new setup for laser cutting	322	The Tool tab	331
The Setup tab	323	The Geometry tab	332
The Stock tab	324	The Heights tab	335
The Post Process tab	325	Simulation results	336
Creating a new cutting tool	326	Summary	337
The Cutter tab	328		

Part 4 – Using Fusion 360 for Additive Manufacturing   339

14

Getting Started with Additive Manufacturing   341

Technical requirements	342	Comparing different 3D-printing technologies	348
Introducing additive manufacturing	342	Fused deposition modeling	349
Exploring the pros and cons of 3D printing over conventional manufacturing processes	343	Introducing stereolithography	350
		Introducing selective laser sintering	352
The pros of 3D printing	343	Summary	353
The cons of 3D printing	346		

15

Managing the Limitations of FDM Printers   355

Technical requirements	356	Improving our prints using support structures	358
Printing overhang geometries	356	Understanding bed adhesion	360
Facing overhangs (undercuts) in 3D printing	357		

Understanding anisotropies of the
printed part   361

Choosing the first layer placement
and part orientation   362

Summary   364

16

Printing Our First Part   365

Technical requirements   365
Presenting the model   365
Creating a new printing setup   366
Orienting the model onto the build
platform   370
Place parts on platform   371

Minimize Build Height   372
Automatic Orientation   374
Generating the support structures 379
Simulating the toolpath   382
Using the post-processor   384
Summary   386

17

Understanding Advanced Printing Settings   387

Technical requirements   388
Creating a new printing preset   388
Understanding general parameters   390
Understanding extruder parameters   392
Extrusion options   393
Extruder 1 options   394

Understanding shell parameters   395
Understanding infill parameters   398
Understanding print bed adhesion   400
Thermal expansion and shrinkage   401
Print bed adhesion   402

Understanding support material
parameters   404
Understanding speed parameters   405
Speed   407
Acceleration   407
Jerk   408

Understanding tessellation
parameters   408
Summary   411

Part 5 – Testing Our Knowledge   413

18

Quiz   415

Technical requirements	**415**	**Answers**	**428**
Questions	**415**	Turning	428
Turning (from Chapters 1 to 5)	415	Milling	428
Milling (from Chapters 6 to 10)	419	Laser cutting	428
Laser cutting (from Chapters 11 to 13)	423	Additive manufacturing	428
Additive manufacturing (from Chapters 14 to 17)	425	**Summary**	**429**

Index   431

Other Books You May Enjoy   442

Preface

This book is not simply an introduction to the CAM module of Fusion 360. Of course, we will analyze in detail all the options of the program, and by the end, we will be able to create toolpaths to use with a CNC machine, that's for sure. However, the real value of this book is the advice and best practices that come from years of experience in the industrial design and production sectors.

If they want to, anyone can download a piece of CAD software and learn it for themselves (this is how I, and many others, learned it); however, issues such as the choice of construction technology (and the relative tools to be used), the feasibility of manufacturing a component, or the optimization of a production lot are much more difficult things to learn independently. That's why I believe that this book is valuable for any novice designer – it will save you from all the mistakes I learned the hard way!

Who this book is for

This book is for 3D enthusiasts or mechanical designers looking to turn their design ideas into 3D models, and their 3D models into final products. Familiarity with any CAD software or Fusion 360 design module is recommended; the book will then teach you the rest.

What this book covers

In *Chapter 1*, *Getting Started with Turning and Its Tools*, we will approach turning, its machining strategies, and the complex geometry of machining tools.

In *Chapter 2*, *Handling Part Setup for Turning*, we will find out how to create a new turning setup using Fusion 360 CAM.

In *Chapter 3*, *Discovering the Tool Library and Custom Tools*, we will discuss several ways of importing new tools into the built-in tool library.

In *Chapter 4*, *Implementing Our First Turning Operation*, we will create a facing operation to remove a layer of material on the stock front.

In *Chapter 5*, *Discovering More Turning Strategies*, we will implement a wide variety of turning operations to machine a given component.

In *Chapter 6*, *Getting Started with Milling and Its Tools*, we will introduce milling, its underlying theory, its main cutting operations, and the tools typically used.

In *Chapter 7, Optimizing the Shape of Milled Parts to Avoid Design Flaws*, we will review the most typical design issues encountered on milled parts and the countermeasures to fix these problems.

In *Chapter 8, Part Handling and Part Setup for Milling*, we will analyze a complex part and choose its placements and machining.

In *Chapter 9, Implementing Our First Milling Operations*, we will create all the toolpaths to entirely machine our first stock placement.

In *Chapter 10, Machining the Second Placement*, we will complete our milling example by implementing advanced operations such as thread milling and 3D profiling.

In *Chapter 11, Getting Started with Laser Cutting*, we will approach laser cutting, reviewing how it works and its hidden complexities.

In *Chapter 12, Nesting Parts for Laser Cutting*, we will discuss several ways to optimize parts on metal sheets using multiple nesting tools.

In *Chapter 13, Creating Our First Laser Cutting Operation*, we will create a new setup and a new tool to generate a cutting toolpath to be exported to a laser machine.

In *Chapter 14, Getting Started with Additive Manufacturing*, we will approach 3D printing, its pros and cons, and the different technologies it's made up of.

In *Chapter 15, Managing the Limitations of FDM Printers*, we will find out how to solve typical issues encountered when using an FDM printer.

In *Chapter 16, Printing Our First Part*, we will analyze the basic settings needed to orient our part onto a build platform and to generate a G-code program.

In *Chapter 17, Understanding Advanced Printing Settings*, we will analyze all the advanced settings used to fine-tune an FDM printing process and their effects on the generated toolpath.

In *Chapter 18, Quiz*, you will face a simple quiz to test your level of knowledge after reading the book.

To get the most out of this book

Since everything will be explained in the simplest possible way, no particular prior understanding of the manufacturing world is required. However, a bare minimum knowledge of common technical terms related to mechanical engineering, as well as some knowledge of the Fusion 360 design environment or similar CAD software, is required.

Please note that this book assumes you are reading it chronologically – for example, basic topics explained at the beginning will not be explained again later. To get the most out of this book, it is highly suggested not to jump through it.

Software covered in the book	Operating system requirements
Fusion 360 (a trial or paid subscription)	Windows or macOS
CoroPlus (a trial or paid subscription)	Windows, macOS, or a web browser
The Nesting and Fabrication extension (a trial or paid subscription)	Windows or macOS

Download the example code files

You can download the example code files for this book from GitHub at `https://github.com/PacktPublishing/Making-Your-CAM-Journey-Easier-With-Fusion-360`.

Download the color images

We also provide a PDF file that has color images of the screenshots and diagrams used in this book. You can download it here: `https://packt.link/1h01Z`.

Conventions used

There are a number of text conventions used throughout this book.

Bold: Indicates a new term, an important word, or words that you see on screen. For instance, words in menus or dialog boxes appear in **bold**. Here is an example: "Select **Custom tools** from the **Tool library** panel."

> **Tips or important notes**
> Appear like this.

Get in touch

Feedback from our readers is always welcome.

General feedback: If you have questions about any aspect of this book, email us at `customercare@packtpub.com` and mention the book title in the subject of your message.

Errata: Although we have taken every care to ensure the accuracy of our content, mistakes do happen. If you have found a mistake in this book, we would be grateful if you would report this to us. Please visit `www.packtpub.com/support/errata` and fill in the form.

Piracy: If you come across any illegal copies of our works in any form on the internet, we would be grateful if you would provide us with the location address or website name. Please contact us at copyright@packt.com with a link to the material.

If you are interested in becoming an author: If there is a topic that you have expertise in and you are interested in either writing or contributing to a book, please visit authors.packtpub.com.

Share Your Thoughts

Once you've read *Making Your CAM Journey Easier with Fusion 360*, we'd love to hear your thoughts! Scan the QR code below to go straight to the Amazon review page for this book and share your feedback.

https://packt.link/r/1-804-61257-X

Your review is important to us and the tech community and will help us make sure we're delivering excellent quality content.

Download a free PDF copy of this book

Thanks for purchasing this book!

Do you like to read on the go but are unable to carry your print books everywhere?

Is your eBook purchase not compatible with the device of your choice?

Don't worry, now with every Packt book you get a DRM-free PDF version of that book at no cost.

Read anywhere, any place, on any device. Search, copy, and paste code from your favorite technical books directly into your application.

The perks don't stop there, you can get exclusive access to discounts, newsletters, and great free content in your inbox daily

Follow these simple steps to get the benefits:

1. Scan the QR code or visit the link below

https://packt.link/free-ebook/9781804612576

2. Submit your proof of purchase
3. That's it! We'll send your free PDF and other benefits to your email directly

Part 1 – Implementing Turning Operations in Fusion 360

In this first part, we will approach a lathe and its most common turning operations, trying to fully understand the potential and main limitations of this particular type of manufacturing process. After diving straight into turning equations and the underlying theory, we will implement a set of turning strategies to fully machine a simple part. In addition, we will cover important topics such as the tool library and custom tools.

This part includes the following chapters:

- Chapter 1, *Getting Started with Turning and Its Tools*
- Chapter 2, *Handling Part Setup for Turning*
- Chapter 3, *Discovering the Tool Library and Custom Tools*
- Chapter 4, *Implementing Our First Turning Operation*
- Chapter 5, *Discovering More Turning Strategies*

1

Getting Started with Turning and Its Tools

Moving from 3D **computer-aided design** (**CAD**) models to manufactured parts may look like a giant leap difficult to overcome. In reality, it is not something to be scared of, and in the following pages of this book, we will try to explain step by step all the major challenges to overcome. In the end, you should have gained more confidence with machining and the manufacturing world in general.

In this chapter, we will approach a lathe for the first time, exploring components, terminology, and best practices without relying on prior knowledge.

The goal of this chapter is to let you familiarize yourself with the theory before approaching real case scenarios. Although you may want to start spinning your chuck immediately, a dive into basic concepts is very important; after all, behind every successfully machined part, however complex it may be, there is always the fundamental underlying theory.

We will also discuss turning, reaching a general understanding of what is possible and what is not possible with this method. This type of knowledge is fundamental to unleash turning's potential, instead of trying to copy and paste provided examples into your machining scenario.

In this chapter, we will cover the following topics:

- Approaching a lathe and its components
- Understanding the main parameters
- Exploring main machining strategies
- Understanding tool geometry

Technical requirements

There are no technical requirements for this chapter; however, just in case you have tools and inserts floating around the room, you may want to collect them in order to observe them as you read.

Approaching a lathe and its components

Turning, long story short, it is a very old machining technique discovered thousands of years ago. While understanding Egyptian pottery may sound interesting to some, it is definitely out of the scope of this book, so we are going to jump right into the action!

Turning is a mechanical process where a cylindrical part is put on fast rotation, and then approached by a cutting tool that progressively removes material from it. The machine that lets us shape our part with this technique is called a **lathe**.

If you ask a professional CNC user, they will tell you that a lathe is a rather complex machine that consists of many components such as the saddle, the tailstock, the headstock, and so on. Don't worry about all these intimidating words—we will approach turning in the simplest way possible. In the following screenshot, we can find a lathe stripped up to the bones:

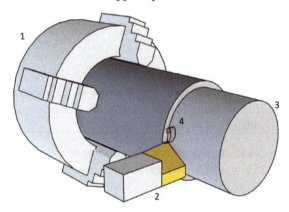

Figure 1.1: Lathe main actors

Ultimately, there are only four actors involved in turning:

1. **Chuck**: A fast-spinning clamping device; it holds the stock to be machined
2. **Cutting tool**: A special hard-metal blade that removes material from the stock
3. **Machined stock**: Our part before machining is completed
4. **Chip**: The removed particles and filament (waste material)

If you have ever approached a working lathe, I'm sure you noticed that the material shape about to be machined—the stock—spins very fast, while the cutting tool moves quite slowly, with a lot of chips being generated and projected all over the surrounding environment.

Even if the chip is valueless, it doesn't mean that we can pretend it doesn't exist; if we did, it would very easily render our part and our tool valueless. Controlling chip formation is very important (not only for turning) because it has the bad habit of becoming an entangled mess (a little bit like pasta) and damages everything it comes into contact with.

There are several types of lathes on the market; some are really big and some are very complex, but at the end of the day, all of them are pretty similar to one another.

Now that we understand the main components of the lathe, we can dive just a little bit into the theory behind such an incredible machine!

Cylindrical coordinate system

In order to understand coordinates, we need to remember that we live in a 3D world. In our world, in order to specify the location of an object with exact precision, we must provide a set of three numbers (coordinates) that measure the distance from a point in space—the origin—in a given direction.

In our daily lives, we often use **Cartesian coordinates**; here, we have a few examples:

- When working in an Excel file or when playing *Battleship*, in order to locate a cell, we need to know the row number and column number (2D Cartesian coordinates)

- If we want to give the dimensions of an object, we must provide the length, the width, and the height (3D Cartesian coordinates)

However, as mentioned, all the examples previously listed are Cartesian coordinates, where coordinates are expressed as a length along axes oriented at 90° from each other.

Orienting all the axes at 90° is generally the best approach for parts with a cubic shape, however, when it comes to lathes, a different coordinate system may be better. This is because all parts machined with a lathe have an axial symmetry, meaning that basically every part looks more or less like a cylinder. Therefore, we can use **cylindrical coordinates** instead.

With cylindrical coordinates, we do not express a point position with three distances; we express a position with an angle and two distances. We can see this in the following screenshot:

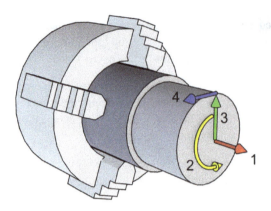

Figure 1.2: Cylindrical coordinates' components

Let's specify what the labels mean:

1.   The red arrow is the axial direction (longitudinal), which is the rotation axis.

2.   The yellow arrow is the rotation angle (α).

3.   The green arrow is the radial direction. Note that the arrow is not static; you can pretend it is a hand of a clock, where all the ticks on the dial would be a possible radial direction. There are infinite green arrows around the rotation axis, each rotated by an angle (α) from 0° to 360°!

      With these three numbers, we already can specify with extreme precision every single point inside and outside our stock. However, it will be useful to introduce another direction.

4.   The blue arrow is the tangential direction. It may not be so simple to understand the tangential direction, but we can just say that it is a vector that starts from the tip of a radial vector with a direction perpendicular both to the radial vector and to the axial vector.

Now that we know a bit more about cylindrical coordinates, we can start using this knowledge to differentiate between turning operations!

Different types of turning operations

To discuss the different types of turning operations, we will look at the two main families of machining:

- External machining
- Internal machining

Let's take a look at these in more detail.

External machining

These operations are the most common operations performed while turning. In this type of process, the material is removed from "outside-in" along the radial direction. This basically means that the average radial coordinate of our tool will keep reducing during the operation.

In the following screenshot, we have a typical external machining operation depicted:

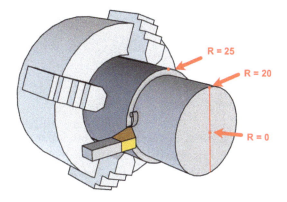

Figure 1.3: External machining example

Just by glancing at the figure, we should understand that the tool starts on the outside of our part where the radial coordinate is at the maximum, while machining it will progressively reduce it when moving toward the center (where the radial coordinate is 0).

Since, as we will discover, there are many different types of machining strategies, it is a bit difficult to find a strict definition for external machining; however, everything will be clearer when compared to the next machining approach.

Internal machining

These operations are different from the previous case since the material is removed "inside-out." So, the average radial coordinate for our tool is growing during the operation:

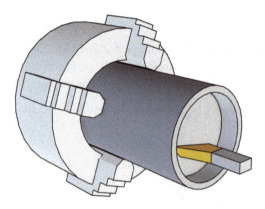

Figure 1.4: Internal machining example

As you can see from the previous screenshot, the tool is moving inside our stock, and the cut chip may find itself stuck in the cavity since there is material all around the tool. As you can also see, the tool is still quite visible, but there are cases where we may completely lose the line of sight after a certain depth, and this is not good for beginners.

As a matter of fact, when we first approach turning, it is always a good idea to look constantly at our part while being machined since we can check the surface finish, prevent unforeseen collisions with parts of the stock or parts of the lathe, and monitor chip evacuation. For all these reasons, internal machining may be a bit trickier than external machining.

> **Note**
>
> As a rule of thumb, if the forming chip is somehow free to flow and falls on the ground, we are in external machining; if it can be stuck inside, we are in internal machining.

It is now time to explore something a bit more technical but really important to understand: working parameters!

Understanding the main parameters

It is now time to talk about the parameters involved in our turning operations. First of all, what is a parameter? In short, a parameter is a value or a setting that we can change.

Of course, some parameters are easier to change than others. For example, changing the axial movement speed for our tool is very simple, while on the other hand, changing the maximum cutting power may require a bigger and more powerful lathe.

Also, as we will shortly discover, some values are somehow connected to others; changing one parameter may change one or more others, so we have to optimize parameters according to our lathe specs and to the part we want to machine.

Let's find out what the main parameters are that we have to work with.

Turning speed

The **turning speed** is a measure of how fast the chuck is spinning; it may be measured in radians per time unit ($\dot{\alpha}$) or in revolutions per time unit (n).

As shown in the screenshot, the bent vector represents the part spinning along the longitudinal axis:

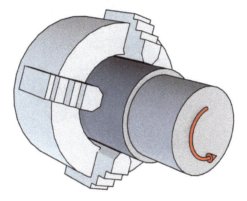

Figure 1.5: Turning-speed visualization

One revolution of our chuck corresponds to an angle (α) of 360° or 2π (depending on whether it's measured in degrees or radians). But, since n is measured in revolution per time unit, we need to divide the angle by the same time unit, so that from the angle, we get angular speed ($\dot{\alpha}$).

In short, we just need to remember this formula:

$$\dot{\alpha} = 2\pi n$$

Old lathes had turning speeds that were only adjustable by manually changing the belt and pulleys, or changing the gears. Today, there are lathes that can control the rotation speed with inverters and encoders. So, depending on your lathe, the turning speed may or may not be a parameter that's easy to change.

> **Note**
>
> Please be careful when dealing with time units: sometimes the time unit may be seconds and sometimes minutes. As you can imagine, one revolution per minute is very different from one revolution per second, so when using any formula, please always be sure of its units!

Cutting speed

The **cutting speed** (V_c) is the relative speed between our cutting tool and the surface it is cutting. There is only one important thing to notice, which is that the cutting speed is highly dependent on the radial position of our tool: the bigger the diameter means the higher the cutting speed (at a constant rotation speed).

As you can see in the screenshot, we have a common rotation speed for the two tools (since the chuck rotation is always the same); however, as you may have noticed from the different arrow lengths, the tool that is machining a bigger diameter has a much higher cutting speed (vertical vector) than the tool machining a smaller diameter:

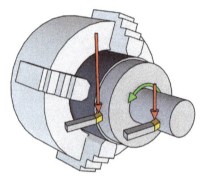

Figure 1.6: Cutting-speed visualization

> **Note**
>
> A vector can be defined by a magnitude and a direction. Magnitude is represented by the arrow length, while the direction is represented by the direction of the pointed tip.

Cutting speed is typically measured in **meters per minute (m/min)**, and there is a very simple formula for calculating it:

$$V_c = \frac{\pi n D}{1000}$$

Here, D is the diameter in **millimeters (mm)** and n is **revolutions per minute (RPM)**.

For example, let's imagine we have a chuck rotating at 500 RPM, and we are performing a longitudinal operation at a constant radial distance of 50 mm (basically, we are machining a stock with a diameter of 100 mm). What is the cutting speed seen by our tool? It's very simple:

$$V_c = \frac{\pi * 500\,[rpm] * 100\,[mm]}{1000} = 157\,m/min$$

Now that we have seen the cutting-speed formula, it should be now clear that when machining a shape with different radial coordinates assumed by our tool, the cutting speed will be subject to changes. Since the working diameter is not a parameter we can change (it is related to the part shape we want to machine), in order to change the cutting speed, we can only adjust the chuck RPM. That's why most of the time, you will see bigger parts spinning slower and smaller parts spinning faster! Once the tool is almost at the rotation axis, the cutting speed will always drop to zero, independently of how fast the chuck is rotating.

Cutting depth

The **cutting depth** (a_p) is a parameter that gives us a measure of how much we impose an interpenetration between our tool and our part; it is measured in mm. Basically, it is related to how much of our cutting edge is engaged with the stock:

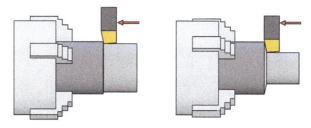

Figure 1.7: Cutting-depth visualization

As we can see, there are two tools machining the same diameter. The one on the left, however, is engaging the stock at a very small cutting depth, and therefore it removes only a thin layer of material. The tool on the right is cutting at a much bigger cutting depth and therefore it is removing a much thicker layer, and most of the cutting edge is engaged.

Please note that cutting depth is not always measured in the radial direction; sometimes, it can be measured in the axial direction. It mainly depends on the type of machining process we use (we will explain this in the *Exploring main machining strategies* section later in the chapter).

Higher cutting depth means more material removed and therefore faster machining, but requires more cutting power and leads to higher stress on our tool and our part. For this reason, please consider that higher cutting depths may partially bend our stock (especially if it is long and supported on one side only). Bending may lead to weird shapes with a rough finish.

Cutting feed

The **cutting feed** (f_n) is a parameter that measures how much our tool is advancing along the cutting direction at every revolution of our chuck, and therefore it is measured in **mm per revolution** (**mm/rev**)

Similar to cutting depth, higher feeds lead to higher machining speed at the cost of higher tool wear and higher power being required. The part finish is highly connected with feed values; as a rule of thumb, a lower feed value means higher surface quality.

In the following screenshot, we have two tools machining the same stock at the same cutting depth:

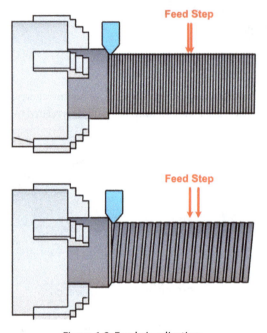

Figure 1.8: Feed visualization

However, in the first example, there is a much smaller feed step than in the second, so though they have the same turning speed, the tool on the bottom will reach the end of the stock much quicker, at the cost of higher surface roughness.

Be careful, as certain operations must be performed at a constant feed, such as threading. Threads have a constant pitch per rotation; therefore, we must set the feed to be equal to the thread pitch in such a scenario.

> **Note**
>
> Threading is the process of making a thread. It may be related to a male thread (such as a screw) or a female thread (such as a nut).

Adjusting the feed will also affect the chip. We can assume that a lower feed will produce a very long and entangled chip, while a higher feed will produce small particles. This is illustrated in the following screenshot:

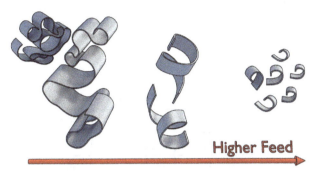

Figure 1.9: Chip formation related to feed

We should always try to hit a sweet spot in chip thickness, width, and length. The ideal chip shape is the one in the middle of the screenshot.

> **Note**
> Chip formation is a very interesting world of its own. I don't really want to distract you from the main topic; however, if you want to learn more about chip formation and different chip types, I suggest you take a look at this interesting link: `https://www.bdeinc.com/blog/types-of-chips-formed-during-cnc-milling/`.

Cutting power

The **cutting power** is the mechanical power required to machine our part at a given set of parameters. It is measured in **kilowatts (kW)**.

Before setting all our parameters, we must always check whether our lathe is powerful enough to handle the machining we want to perform. This is the simple formula to remember:

$$P_c = \frac{V_c * a_p * f_n * K_c}{6 * 10^4}$$

Here, P_c is the cutting power measured in kW; V_c is the cutting speed measured in m/min; a_p is the cutting depth measured in mm; f_n is the feed step measured in mm/rev; and, finally, K_c is the specific cutting force measured in **MegaPascal (MPa)**.

As you may have noticed, there is a new value called K_c that we haven't seen yet. K_c is a parameter related to material strength, tool shape, feed ratio, and cutting depth. This is a difficult parameter to evaluate, and I don't want to bother you with complex calculations… but luckily, I don't have to. We can simply refer to tables where equations are already sorted out by experts. Please note that this is a rough approximation of the real value, so you may want to take it with a grain of salt and have a bit of a safety margin on cutting power!

Here, we can find an example of the most common iron alloys:

Material/Feed	0.1 mm/rev	0.2 mm/rev	0.3 mm/rev	0.4 mm/rev	0.6 mm/rev
Cast iron	3200 MPa	2800 MPa	2600 MPa	2500 MPa	2300 MPa
Mild steel	3600 MPa	3100 MPa	2700 MPa	2500 MPa	2300 MPa
Medium steel	3100 MPa	2700 MPa	2600 MPa	2500 MPa	2300 MPa
Hard steel	4050 MPa	3600 MPa	3300 MPa	3000 MPa	2600 MPa
Tool steel	3200 MPa	2900 MPa	2600 MPa	2500 MPa	2400 MPa

Figure 1.10: Kc approximation for steel and iron

Using such a table is quite simple: we simply select the row according to the material we are about to machine, then we select the column with a feed similar to the one we plan to use, and the result approximates the real K_c value.

> **Note**
>
> If you are intrigued by this parameter, you can find all the formulas for a precise evaluation of K_c here: http://www.mitsubishicarbide.net/contents/mhg/enuk/html/product/technical_information/information/formula4.html.
>
> As you can see, there is a very useful formula where we can insert our cutting parameters to get our K_c value. If you are hungry for more formulas, there is a very good explanation about K_c here: https://www.machiningdoctor.com/glossary/specific-cutting-force-kc-kc1/#k1c-and-mc-chart-for-material-group.

With that last equation, it is now easy to understand that all parameters are connected and that we must change them accordingly to meet the maximum cutting power available for our lathe.

You may be wondering whether it is a must to target maximum power. No, it is not, but from the production point of view, not using our machine at its full potential will cost us more money per hour. However, a typical scenario where we will not use maximum cutting power is when we are machining a fragile material that we may damage with our chuck closed at maximum force to sustain chuck torque.

Now that we have covered the main parameters, we should better understand our lathe. Next, let's jump into something juicier: turning operations and strategies!

Exploring main machining strategies

When we plan how to machine our part out of the stock, we have to think like a chess master. What I mean by this is that we have several pieces we can play with, but in order to win the game, we have to use every component in the best possible way it can be used. Therefore, we need to plan which piece to play first and with which type of move: do we move the bishop or the queen? Do we move by one square at a time or full speed ahead?

In the following sections, we are going to discover the main moves we can play with when it comes to turning.

Longitudinal operations

Longitudinal machining is a simple and common operation strategy, suitable for rough machining with high power and strong tools. Let's check a basic example in the following screenshot:

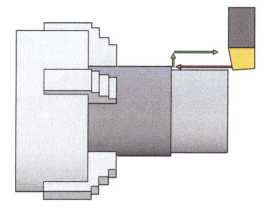

Figure 1.11: Longitudinal machining approach

The typical approach for longitudinal operations is to move the tool on the front of our stock (at a safe distance), set the cutting depth in the radial direction, and then move the tool forward. As we can see in the screenshot, while cutting, our tool is moving along the axial direction only, so we have a constant cutting speed per cutting pass. Once at the desired distance, the tool will disengage our stock radially and then repeat the process until the right diameter is machined.

Since this is longitudinal machining, the main cutting direction is on the longitudinal axis at a given radial distance; therefore, after a chuck rotation, our tool will be moved by one feed step along the longitudinal direction.

As a recap, in this type of machining, the cutting feed is along the axial direction, while the cutting depth is fixed and constant and is measured along the radial direction. As we are about to discover, there are other machining strategies where what we just said doesn't apply.

Facing operations

Facing is used when we need to clean the front face of our stock:

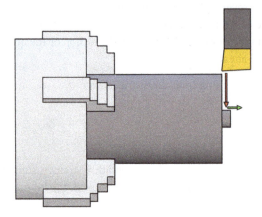

Figure 1.12: Facing machining approach

As shown here, while cutting, our tool is moving in the radial direction only; therefore, the cutting power will keep changing unless we adjust the feed or rotation speed accordingly.

The standard approach for facing operations is to move the tool at a safe distance in the radial direction, set the cutting depth in the axial direction, and then approach the stock radially. Once at the desired position, the tool will disengage our stock axially and then repeat the process until the right length is machined.

Since in facing, we are basically cleaning a slice of our stock, we are operating at a fixed axial coordinate and our tool is moving radially. So, after a chuck rotation, our tool will be moved forward by one feed step along the radial direction.

As a recap, when facing, the cutting feed is along the radial direction while the cutting depth is along the axial direction (this is different than longitudinal operations!).

Plunging operations

Plunging is similar to facing since the tool is moving in the radial direction only; however, because plunging is used for grooving or cutting, the tool is much slimmer. That's because circlips grooves can be very thin, and our tool needs to fit inside them. Another reason is that when cutting our part, it is not a good idea to perform a large cut because that would require a lot of material removed without a real need.

As already mentioned for facing, since radial coordinates are always changing, so does the cutting power, unless controlled by feed changes or RPM.

As shown in the following screenshot, the typical approach for plunging is to position our tool at a safe radial position, set the desired axial coordinate, and then approach the part radially:

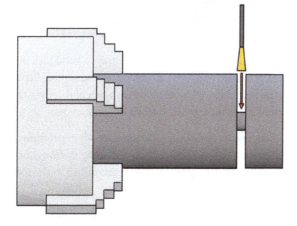

Figure 1.13: Plunging machining approach

So, to recap this strategy, plunging is similar to facing since it has feed along the radial direction too, but the main difference is that we cannot set the cutting depth since the tool cutting edge is always engaged entirely; therefore, the cutting depth is always at the maximum possible value!

Profiling operations

Profiling is one of the most complex operations since the tool moves constantly radially and axially. Let's check the following screenshot for an idea of how profiling works:

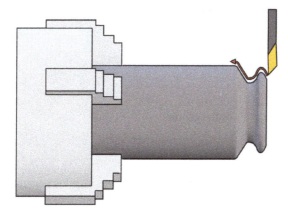

Figure 1.14: Profiling machining approach

As shown in the screenshot, the motion path of our tool can be very complex; therefore, this type of operation is the most flexible and is well suited for finishing our part. However, due to the fact that the cutting direction is not constant, there is always the risk of impact between the tool and our stock. As you may have noticed in the screenshot, after "climbing the first hill," our tool will likely collide with the stock since the required path is too steep for the tool to pass.

To reduce the possibility of errors, an impact simulation is always a good idea in CAM, but for complex machining operations such as profiling, you must always check the results. Especially when a beginner is approaching turning for the first time, 99% of the time they will discover that the back of the tool or the shank will collide where the profile gets too steep or too narrow.

If we want to analyze what happens to our cutting parameters when profiling, we have to remember that the profile is not constant along the radial direction nor along the longitudinal direction; it keeps changing. So, unless our chuck is able to control rotation according to the tool's radial position, the cutting speed will not be constant.

Having said that, we also need to remember that feed is measured along the cutting direction, so it will keep changing along the profile; the same can also be said for cutting depth.

Finally, we are at the end of yet another section. There is only one last major topic to cover: tool geometry.

Understanding tool geometry

As we discovered, there are many possible operations and geometries that we can machine on our lathe, but not every tool is suitable for every operation (always remember the chess analogy). We are now going to classify tool geometry in order to recognize, at first glance, which tool is more suitable for which machining strategy.

We can split the geometry of a cutting tool into three main categories:

- Tool edges
- Tool surfaces
- Tool angles

Let's start with our deeper analysis of the geometry.

Tool edges

In the following screenshot, we can see a tool with a transparent shank where the edges are highlighted:

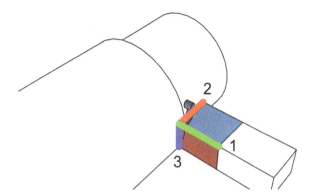

Figure 1.15: Main tool edges highlighted

You'll notice that the tool shown is very crudely drawn and is simply a square. In reality, as we are going to discover, tool geometry is way more complex; however, at the moment, we will pretend it is a real tool since understanding surfaces and edges will be simpler this way.

Here are the three edges:

1. The main cutting edge
2. The auxiliary cutting edge
3. The nose edge

Let's take a look at them.

Cutting edge

As the name suggests, the **cutting edge** is a sharp edge that cuts our stock that is rotating against it. The cutting edge simply cuts along the feed direction and cuts the main portion of the chip.

Auxiliary cutting edge

In order to cut the stock, we need to cut in two directions. So, the **auxiliary cutting edge** cuts the side of the chip and removes the material from the stock.

Nose edge (radius)

The **nose edge** is not a real "edge." It is never sharp because it wouldn't last a single second while machining; instead, it is a "smooth edge" with a specific radius that can be bigger or smaller according to the tool we choose. A tool with a bigger nose radius will be less sharp and therefore will be able to sustain higher feed rates and higher cutting depths for a longer time.

Saying that, it is important to note that the nose radius is responsible for the surface's finish. As you can see here, from a geometrical point of view, a higher nose radius will lead to less surface roughness when machining, and therefore to a better surface finish:

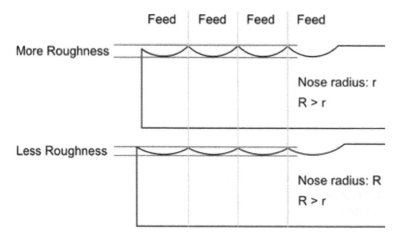

Figure 1.16: Changing the nose radius will change surface roughness

So, it looks like you may always want to use a large nose radius for roughing and for finishing… well, not so fast—there is a catch! When working at a low cutting depth (for example, when finishing a flexible part), a big nose radius will generate higher radial forces that are going to create strong vibrations (more details on that in the *Entering angle (KAPR)* section at the end of this chapter). Such vibrations will ruin the surface finish. A tool with a small nose radius will be super sharp and short-lasting, but also incredibly more effective at small cutting depths when finishing.

Now that we have learned something about the three main edges of every tool, let's now start talking about the three main surfaces and their relative orientation.

Tool surfaces

All the edges listed up until now come into contact at the tip (or the **nose**) of the part. Three main surfaces are defined from this intersection. As always, it is way easier to understand the geometry by looking at a simple example:

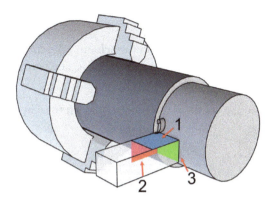

Figure 1.17: The main surfaces of a tool

Here, we can see three highlighted surfaces to study:

1. Rake surface

2. Main flank

3. Auxiliary flank

Let's take a look at them.

Rake surface

The **rake surface** is the surface generated between the main cutting edge and the auxiliary cutting edge. It is the surface our chip is rubbing against before being evacuated. As you can imagine, much of the cutting power is transmitted to the stock through the cutting edge and the rake surface. At the moment, there is not much more to say about it, but as we will shortly discover, the main point of interest is its rotation.

Tool flanks

In our tool, we have two flanks:

*   The **main flank** is generated between our cutting edge and our nose edge

*   The **auxiliary flank** is generated between our auxiliary cutting edge and our nose edge

After that quick look at surfaces, let's analyze angles.

Tool angles

When machining our part, we have many angles involved; some angles are related to the tool shape, while others are related to the way our tool is installed on the lathe.

Rake angle

The **rake angle** (γ) is the angle that describes how much our rake surface is rotated around the cutting edge. We can classify tools according to their rake angle, and we can distinguish three types of tool families: negative tools, neutral tools, and positive tools.

In the following screenshot, we can see those three tools, from left to right respectively:

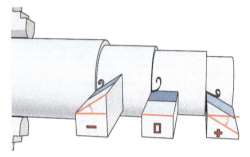

Figure 1.18: Rake angle visualized

More positive values will result in a super sharp tool (with a lower life span) capable of an easy-flowing chip very suitable for a surface finish.

More negative values will result in a stronger tool with a higher life span. Since the cutting edge is not so sharp, the tool will be great for roughing passes with higher cutting depth. As you can see in the screenshot, negative rake angles impose a hard twist in the forming chip; therefore, it won't flow so easily and will break into smaller particles that may rub against the machined surface.

There is not much to say about neutral tools—their properties are a mix between positive tools and negative tools.

Please note that there is also a less important rake angle measured from the auxiliary cutting edge:

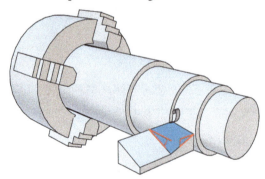

Figure 1.19: Rake angles

In the screenshot, we can find the main rake angle on the left and the auxiliary rake angle on the right.

Relief angle

When introducing tool flanks, we missed one important thing—they should never rub or collide against our stock or our machined part.

As you can imagine, the only part of our tool that should ever rub against our stock is our cutting edge(s). In the tools seen up until now, there is actually a really large surface rubbing against the machined part. Let's analyze the following screenshot for a better understanding of what is going on:

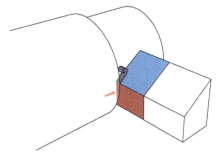

Figure 1.20: Friction area for a squared tool

As shown in the screenshot by the arrow, all that area is rubbing against the machined side! Not good at all! The same is true on the other flank.

To prevent this type of behavior, in real life, the two flanks are always rotated along the cutting edges by a **relief angle** in order to avoid friction. Every tool always has relief angles, and the angle value cannot be negative; otherwise, cutting edges would never engage the stock (because the bottom of the flank will hit the stock first):

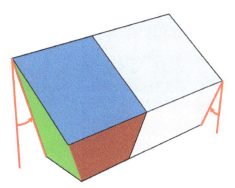

Figure 1.21: Relief angles displayed

We can find in the previous screenshot the auxiliary relief angle (on the left) and the main relief angle (on the right). These angles avoid friction with the turning stock while machining.

Nose angle

While looking at all the tools displayed up until now, you may have noticed that the angle between our cutting edge and our auxiliary cutting edge—called the **nose angle**—is constantly 90°. Just in case you are wondering… more often than not, this is not true; there are many different shapes available on the market!

Let's check the following screenshot to grasp what the nose angle is:

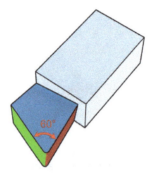

Figure 1.22: Nose angle visualized

Why don't we try to analyze the tool shown in the screenshot? We should have lots of ways to analyze it. At the first glance, we should recognize that it is a positive tool (but almost neutral), that it has small relief angles on the two flanks, and as measured by the arrow, the nose angle is 60° with a small nose radius as well. Not bad!

This is a somewhat old type of tool; it has a large drawback: once the cutting edge or rake surface is really damaged, we have to throw it in the bin and buy a new tool. Today, the most typical approach is to have a tool composed of a shank and a swappable insert that can be replaced without changing the entire tool.

In the following screenshot, we can see on the left we have an "old-style" tool made in a single piece, while on the right side, we have a tool shank and a replaceable insert (both have the same cutting geometry):

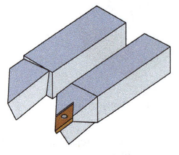

Figure 1.23: Insert versus solid tool

Also, please note that most of these inserts are symmetrical, so they have multiple cutting edges; once one of them is spent, we simply rotate the same insert and keep spinning the chuck!

In the following table, you will find a few typical insert examples with different nose angles and their suggested use:

TOOL SHAPE							
Name	R (Round)	S (Square)	C (Rhombic 80°)	W (Trigon)	T (Triangle)	D (Rhombic 55°)	V (Rhombic 35°)
Preview	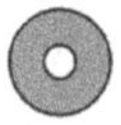						
Nose angle	0/90°	90°	80°	80°	60°	55°	35°
TOOL USAGE							
Versatility	0	0	+++	++	++	+++	+
High-depth roughing	+++	+++	++	+	+	0	0
Low-depth roughing	0	+	++	+++	++	+	0
Finishing	0	0	+	++	+++	+++	+++
Profiling	0	0	0	+	++	+++	+++

Figure 1.24: Tool specifications and typical use

The table shown is divided into two sections. The first part is related to the tool shape, where we can find the seven different families, their naming convention, and their shape. The second part of the table rates tool usage from "+++" to "0" (with "++" and "+" values in the middle). Let's review the usages:

- **Versatility**: If we were alone on a desert island with our lathe, which tool would we take with us? A higher ranking ("+++") means that that tool is a good choice in many situations and can more or less do everything. It is a workhorse and not choosy!

- **High-depth roughing**: If we need to remove a lot of material with rough passes at high cutting depth and high power, we need a very strong tool. A higher ranking ("+++") means that the tool is tough and long-lasting, while a lower ranking ("0") means that the tool is super sharp and will likely break or wear immediately.

- **Low-depth roughing**: In this scenario, we still need to perform roughing passes but with a thinner cutting depth. This case is somewhere between a roughing pass (where we remove a lot of material and we don't mind about tolerances and surface quality) and a finishing pass (where we try to get good surface quality and tight tolerances). I decided to include it just to let you note that at low cutting depth, the round shape is very bad because it causes vibrations (this will be described later in the *Entering angle (KAPR)* section).

- **Finishing**: If we need to finish our part, we probably need to remove a thin layer of material. A higher ranking ("+++") means a super sharp tool suitable for a good surface result.

- **Profiling**: With this ranking, we are going to evaluate which tool is better for machining complex profiles. A higher ranking means that the tool is very good and that it is not limited by accessibility issues, while a lower ranking means that the tool may often collide with the stock. The main difference between profiling and finishing is that the profiling score also considers shape accessibility.

We just mentioned shape accessibility—what does that mean? The answer is simple: slim and sharp tools can reach areas beefier and stronger tools cannot. Let's check the following screenshot:

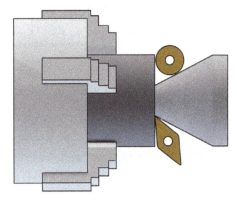

Figure 1.25: Different accessibility displayed

As we notice at the first glance, the round tool (type *R*) cannot remove as much material as a rhombic tool (type *D*) because the area to be machined is too deep to reach for it.

We have just taken a look at all the angles that we can find when looking at a tool on the workshop shelf. Is this the end? Not really—despite the tool shape, we can always decide to mount it rotated! Let me introduce you to the entering angle.

Entering angle (KAPR)

The **entering angle** is the angle measured from the feed direction to the cutting edge.

Changing the entering angle can dramatically change tool performance and chip management. In normal circumstances, it is always a good idea to set the KAPR close to 90°. With flexible parts, we should never set an entering angle lower than 70° if possible. The reason is pretty simple: changing the angle is actually changing the way our cutting edge is transferring power to the stock material:

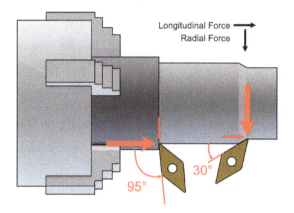

Figure 1.26: The same insert mounted at different KAPR angles

As shown on the left of the screenshot, when close to 90°, our tool is transferring power mainly in the longitudinal direction, and only a small percentage of the total is transferred as the radial component.

On the right, we can see that the tool is far from 90°, and therefore the radial component is much higher; this is bad, especially with flexible parts, because they will bend.

Why is bending a bad thing? Let's review the following screenshot for a better understanding:

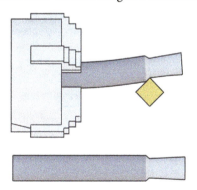

Figure 1.27: Bending

As you can see, the screenshot depicts a scenario where a slim part is machined with a longitudinal operation. Other than that, we can see that the tool has a 45° entering angle and a high cutting depth, which is the perfect recipe for high radial forces that will bend the part while machining.

As you can see at the top of the screenshot, the stock is heavily bent while machining, so we can imagine high vibrations due to the runout of the center rotation line and low surface quality.

As we can find at the bottom of the screenshot, after the machining is completed, what should have resulted in a nice cylindrical surface appears more like a cone with a higher diameter where the bending was more severe!

It may look like a small entering angle is forbidden, but not so fast. There are many different scenarios where we may have to use small angles; here are some examples:

- **When we want to improve tool life span**: At the same cutting depth, with a smaller entering angle, a bigger part of our cutting edge is engaged, and therefore the cutting force will be distributed over a bigger area with smaller stress on the insert that will last longer.

- **When we want to machine a thread**: Let's take a look at the following screenshot for a better understanding of this example:

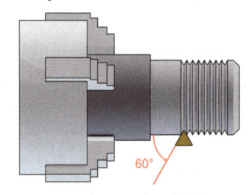

Figure 1.28: Threading operation displayed

As shown in the previous screenshot, when threading we must use an entering angle and a nose angle strictly dictated from the thread specs: 60° as KAPR and 60° as the nose angle (for most metric threads); that's why the triangle insert is very often used for threading.

As we already mentioned, remember that when threading, many parameters are locked to thread specs: from the thread pitch, we must impose a certain feed step, and from the thread angle, we must use a tool with a specific nose angle, entering angle, and maximum nose radius. Do not underestimate the nose radius when threading: if it's too large (over thread specs), the nut will not be able to be screwed on.

So, when threading, if we need to reduce vibrations or bending, there is only one parameter left… we can change the cutting depth only.

As a bonus point, let's explain why the round tool shape can be a good choice at high cutting depth but may generate high vibrations at low cutting depth. Let's analyze the following screenshot:

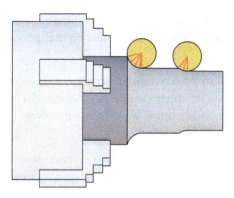

Figure 1.29: Round-tool KAPR

As you can see, the tool on the left is cutting at a high cutting depth; therefore, it is exchanging all the cutting force on a wide range of angles, from almost 90° to 0°, so the radial force generated is not a main issue. On the other hand, the tool on the right is engaging a thin layer only, and the cutting force is shared from almost 0° to 0°. Such a small KAPR will create massive vibrations—that's why round tools are good for high-depth machining but not so much for low depths (and the same can be said for the nose radius).

When facing, a round tool may become handy even at low cutting depth since we don't care that much if the force is transferred to the chuck!

There are many more formulas related to the entering angle, but they are probably way beyond the scope of this book, so we won't deal with them. A last tip on the subject, though: just note that the entering angle is also responsible for the chip shape, so remember that the smaller the entering angle, the thinner and the wider the chip (and vice versa).

As shown in this last section, we should always pay close attention to the shape of the tools since they may look similar, but in reality, they are very different and may be the cause (or the solution) of many machining-related problems.

Summary

Congratulations! That was the end of the first chapter—I hope that the journey up until now was easy and pleasant.

Let's quickly recap what we learned during this chapter. Firstly, we saw what turning is and how it shapes our parts, and discovered what the main components of a lathe are. After this, we moved to something a bit more technical: studying the cutting parameters and how we can control them and evaluate the cutting power required. Then, we moved forward to the main machining strategies used when turning and saw the main geometries of our tools and how they can affect machining.

All these elements are key for better awareness while planning a machining strategy. I encourage you to experiment and widen your knowledge by studying the specs of the tools you use on the manufacturer's datasheet.

Please note that this chapter was an introduction to turning. We should now understand a bit better what we are talking about, but we definitely cannot consider ourselves turning experts. We had to simplify many concepts, and we made a brutal approximation; however, for most typical beginners, the provided knowledge level should be spot on.

Now, follow me to the next chapter where we are going to experiment with something more practical: using the Fusion 360 CAM module for our first turning machining setup.

2

Handling Part Setup for Turning

In this chapter, we will run the Fusion 360 **Manufacture** environment for the first time and learn how to approach our part setup.

Mastering the **Setup** panels is not very simple because there are many different tabs, commands, and parameters to understand and tweak. However, the goal of this chapter is to explore in detail (whenever possible) all of the options that the setup process requires us to specify.

Sometimes, we can simply try to copy someone else's parameters and pray that they fit our needs too. Doing so may lead to a fast start, but it will also leave gaps in our theory. The best mindset to have is to always critically watch every option and wonder why it is set the way it is and what would have happened if it were different.

In the following chapters, we are going to have lots of practical examples to analyze and follow, but for now, let's try to arrive at those chapters with a solid background understanding; it will be much more rewarding!

In this chapter, we will cover the following topics:

- Exploring the main interface of a **computer-aided manufacturing (CAM)** project
- Finding example projects in Fusion 360
- Importing our chuck
- Understanding how to set up the first part to be machined

Technical requirements

The only requirement for this chapter is to have a Fusion 360 license (at any level) so that we can launch the CAM module and start using it.

Exploring the main interface of a CAM project

Before jumping into action, let's review the main interface for the CAM module. It is quite similar to the **computer-aided design (CAD)** module interface:

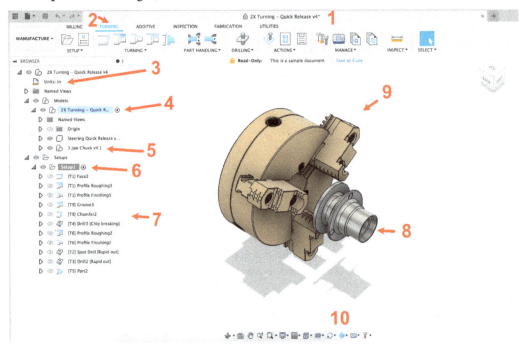

Figure 2.1: Main Fusion 360 interface

Please note that we are inside the **MANUFACTURE** module, which is a fancy way to refer to CAM. All the commands and panels we will use in this chapter are from the **MANUFACTURE** environment, so please be sure to set the working environment correctly.

> **Note**
>
> CAM is a term to refer to software we use to program our **computer numerical control (CNC)** machine. We can find all the CAM commands inside the **MANUFACTURE** environment of Fusion 360.
>
> On the other hand, CAD is a term to refer to a type of software to design our parts. As you should already know, you can find all the CAD tools inside the **DESIGN** environment of Fusion 360.

Let's review the most important elements on the screenshot:

1. **Project name**: In our case, it's **2X Turning – Quick Release**.

2. **Machining tabs**: Under the different tabs, we can find commands related to different machining types. As we can see in the example, the **TURNING** tab is selected, and we can find all the machining strategies such as *facing* and *profiling*.

3. **Project units**: The units can be shown in **millimeters (mm)** or **inches (in)**. We should always be very careful when it comes to which measurement units the project uses; in some examples, it may use mm, while in others, it may use in—therefore, copying numbers or tools from a built-in example to our project may cause confusion.

4. **Main model tree**: Here, we can find all the bodies made in the **DESIGN** environment of Fusion 360.

5. **Chuck model**: As you can see, the chuck is part of the 3D geometry. We can have models to be used just for reference—for example, vises, chucks, build platforms, and so on.

6. **Setup folder**: Here, we will set stock dimensions and the **World Coordinate System (WCS)**. This is the coordinate system that all our turning operations inside this setup will use as the origin, and will be explored later in the chapter.

7. **Operations list**: Here, we can find all of the machining steps required to build our part. For example, in this list, we have all the operations, from the very first roughing passes to finishing passes up until the stock cut once the part is completed.

8. **Part model**: This is the final 3D model that we want to get after all the turning operations are performed on the stock.

9. **Chuck 3D model**: A 3D model used just for reference for collision analysis.

10. **View options**: In the **MANUFACTURE** environment, we have multiple new options (not available in other environments) related to the CAM path and CAM elements. For example, we can display the toolpaths, the machine, the machined stock, and so on. We are going to use these commands in the following chapters, but at the moment, just remember that these options are a bit different from those found inside the **DESIGN** environment that you are used to using.

If you opened the **MANUFACTURE** environment for the first time, the interface will be a blank default. However, there are many built-in examples where we can study someone else's work.

Finding example projects in Fusion 360

One reason why Fusion 360 is a really good entry point for beginners is that there are many examples already built into the program that we can use to start learning from.

Let's find out where those examples are located:

1.   Open the side panel by clicking on the icon that looks like a grid of squares.

2.   With the side panel now open, you can scroll through the projects.

3.   Then, select the folder called **CAM Samples**:

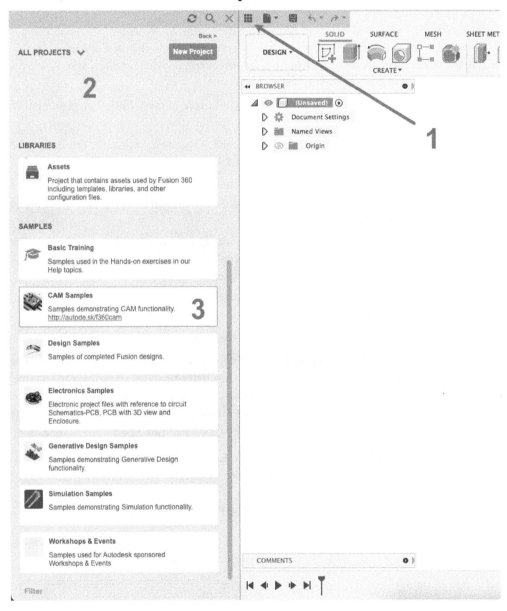

Figure 2.2: Location of the CAM Samples folder

Inside this folder, there are tons of built-in projects related to CAM and machining in general. In fact, the number is so high that it may be a bit overwhelming at first, but over time, you will explore the folders and find examples of interest.

Please also note that the **CAM Samples** folder is not only useful to study machining operations—inside the folder, there is another folder called **Workholding**; here, you can find many useful models that we can use, such as vises, chucks, working tables, clamping devices, and so on.

This is great news since it is always a good idea to check for collisions while programming our CNC. In order to avoid any collision when machining, we should consider the shape of our lathe and all its components close to the machining area such as the chuck and the tool holder; therefore, we can model our lathe and import it inside our CAM project. But we may also find something already pretty close to our machine without wasting a single minute on modeling all the components.

Let's now open the project example named **2X Turning – Quick Release** (located in the **CAM Samples** folder) and jump to the next section.

It is now time to start our CAM project. We are supposed to already have a part to machine, but maybe we don't have a chuck inside the project – let's just import one for our needs.

Importing our chuck

As we already discussed, in order to avoid collisions between our tool and our chuck, it is a good idea to have something that emulates the size of our chuck or our components inside our 3D world. We are about to discover how to simulate chuck placement; however, just before moving forward, I'd like to highlight what I consider a bad type of advertising.

If you have ever watched any advertisement for CAM software, you have probably noticed that alongside the machined part, not only there is a chuck or a tool holder in place, but there are also lots of complex details of the machine. The truth is that all that stuff is there just for marketing purposes!

In order to find possible collisions between the stock and our machine or between the tool and our machine, a set of properly placed boxes is as effective as a super fancy 3D model with all those extra details. However, Fusion 360 provides a pretty big group of work-holding tools, so we may decide to use some of them without an extra modeling effort.

Since the sample projects are in read-only mode, it is a good idea to start from one of the built-in models and save it with another name in our project folder.

There is a very fancy chuck model (in other words, overly detailed) that we are going to use for the first examples, which is called **Generic 3 Jaw Chuck** (located inside **CAM Samples | Workholding**). So, open the **Generic 3 Jaw Chuck** model and then save it in your active project folder. Now, you can open the copy and modify it according to the chuck specs of your machine (remember—you don't really need to create anything super detailed!).

Let's take a look at the sample chuck:

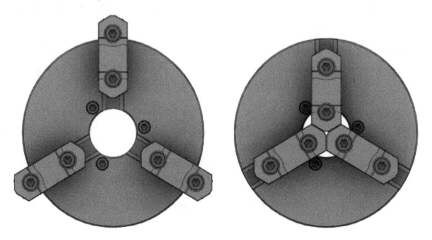

Figure 2.3: Built-in chuck

You will note that this model is very smart, including a set of constraints that can simulate the motion of the **jaws**. As shown in the screenshot, we can click and drag one of the jaws to open or close our chuck (that's pretty much useless, but very impressive nonetheless!).

> **Note**
> The jaws of the chuck are moving parts that close around our stock with a strong grasp and transfer the chuck torque when spinning.

It is now time to import the chuck inside the 3D CAM project. Since this book is not aimed at complete beginners of Fusion 360, we will assume that you are already able to produce complex 3D models inside the **DESIGN** environment; for that reason, we'll suppose that we have already designed our part to be machined and we want to import our chuck inside the assembly.

To do this, it is a very simple procedure:

1. Open the side panel (if closed).
2. Navigate through the project folder.
3. Locate the **Generic 3 Jaw Chuck** file inside the folder.
4. Right-click on the file and choose **Insert into Current Design**:

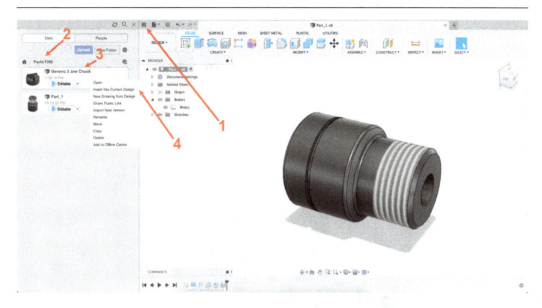

Figure 2.4: How to insert a chuck into our current design

After the import, we should be presented with something like this:

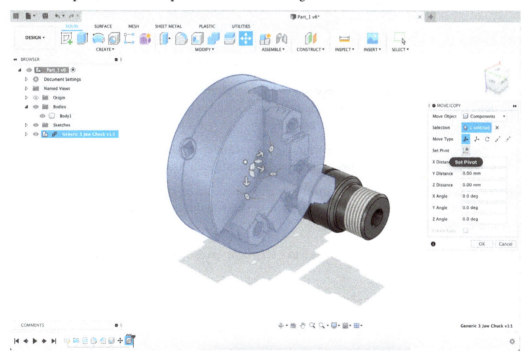

Figure 2.5: Chuck imported

As shown in the screenshot, after we import the chuck, we can decide the position of our 3D model inside the world. We can use all the common tools we should already know and love from the **DESIGN** environment to translate and rotate the chuck model properly, making sure to align the chuck rotation axis with the part rotation axis.

Once we have aligned them, we can move the chuck and set the distance between our machined part and the end of our chuck. But how far from the part do we have to place the chuck?

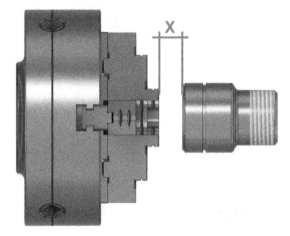

Figure 2.6: Chuck distance

To answer this question, we have to consider a few important points:

- Despite the part length, do we need to machine the entire length or only a certain area? If we don't need to machine the entire length, we may decide to insert a portion of our part inside the chuck; in that case, **X** would be negative.

- Do we want to cut the part after machining? If so, we must consider extra space for tool width, shank size, and insert clearance. In this case, **X** would be positive.

- Is it a flexible part? Remember that the higher the distance, the higher the bending while machining. So, you may want to set it just to the correct distance to be able to cut it and not much more.

In most cases, there is no real reason to work in a very tight space between the rear of the machined part and the end of the chuck.

Now that we have positioned the chuck, we can move forward to our first CAM setup!

Understanding how to set up the first part to be machined

First of all, what is a **setup**? The answer is very simple—it is a set of commands and parameters that we specify for the starting conditions of our project. It is the first mandatory step in order to move forward with any machining. All those conditions and parameters will tell Fusion 360 how to use and optimize the machining process.

> **Note**
>
> We may have a single setup or multiple setups one after the other. We need a new setup every time we plan to change the coordinate system or machining processes. We will cover multi-setup operations in *Chapter 9* and *Chapter 10*.

So, where is the **SETUP** button? If you can't find it, let's check its location:

1. Ensure that you're in the **MANUFACTURE** environment.

2. Then, open the **TURNING** panel.

3. Select the **SETUP** button (which looks like an open folder):

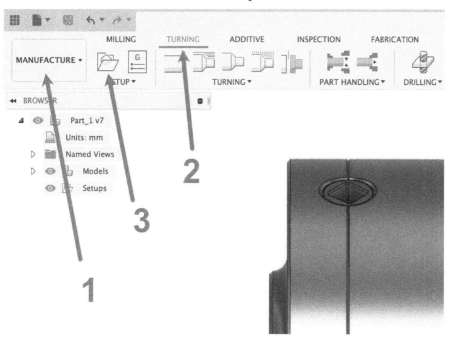

Figure 2.7: The SETUP button

When you launch the command, there are three main tabs:

- **Setup**
- **Stock**
- **Post Process**

Let's take a look at each of these now.

Setup tab

Inside the **Setup** tab, we can specify all information related to the coordinate system and the machine geometry. In the following screenshot, we find a preview of the panel:

Figure 2.8: The Setup panel

Here, there are several subpanels to be aware of:

- **Machine**
- **Setup**
- **Work Coordinate System (WCS)**
- **Safe Z**
- **Model**
- **Chuck**

Let's go through them one by one.

> **Note**
>
> The **Setup** panel and subpanels are going to be influenced by the type of machining we are using, so not every command or option may be available with every setup type. At the moment, we are going to analyze a turning setup.

Machine

This is a super fancy option that allows us to import a built-in CNC machine inside the working project for simulation purposes. At the moment, the Fusion 360 library doesn't come with any lathe models to use; however, don't worry about this—as already discussed, importing a complete 3D machine in our project is cool but also pretty much useless, and the option is not mandatory during setup anyway.

Setup

Inside the **Setup** subpanel, we can find all the options to specify the type of machining we want to implement. Here, we have two options:

- **Operation type**
- **Spindle**

What do they mean?

Operation type

Operation type refers to the machining we want to perform. There are different types of machines out there, and not all of them are lathes; some are milling machines, some laser cutters, and some combine multiple technologies—therefore, we must specify the type of machining we want to use.

Spindle

In the case of a **Turning** setup, we are asked to provide information about chuck configuration. What does this mean? The answer is that not every lathe is similar to others; there are advanced types of lathes with more than a single chuck (also called a **spindle**).

But why would anyone use two chucks instead of just one? There are two main reasons. We have already mentioned one: flexible parts may require additional support in order to be machined without excessive bending. But the second reason is that when using a single chuck, we can only machine one side of our stock since part of it is inserted inside the chuck itself. With two chucks, we can overcome that limitation; check the following screenshot for a better understanding:

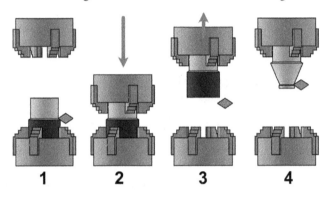

Figure 2.9: Two chucks working

We can split the procedure into four steps:

1. The cutting tool cannot move anymore; it is very close to the main chuck, and most of the stock we need to machine is inside the jaws. In short, the part of the stock we need to machine is inaccessible.

2. The tool is moved at a safe radial distance, and the auxiliary chuck closes the jaws onto the stock already machined.

3. The auxiliary chuck is then moved away from the main chuck.

4. The tool can now start machining the previously hidden side of the stock.

As you can imagine, this type of operation is a bit more advanced, so for the sake of simplicity, we are going to ignore multiple chuck setups (however, do be aware of this useful option).

Work Coordinate System (WCS)

This is one of the most important sections inside the **Setup** panels, but what is our WCS?

WCS is the local coordinate system of the stock we want to machine. Every position or movement of our cutting tool will be described as a set of coordinates measured from such a coordinate system. It is a bit like the *Battleship* game where we need to set a coordinate system to manage ship position and shot positions.

In the same way, we need to tell Fusion 360 where the origin of our coordinate system is and where our directions are pointing. As explained in, we already mentioned that Fusion 360, as with almost every CNC, is working on Cartesian coordinates; this means that after specifying the origin location, we have to set x, y, and z axis orientation.

Fusion 360 is quite capable of automatic stock placement so, most of the time, it is able to preset the coordinate system properly with the right origin and the right rotation axis. That's not always the case, though…

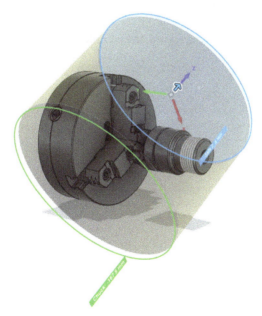

Figure 2.10: Wrong automatic placement

As we can see in the preceding screenshot, with complex shapes or if we imported a chuck or other fixture, it may mistake the WCS placement and orientation. This means that we have to manually specify the coordinates.

How to do that? Simply click on the **Orientation** dropdown:

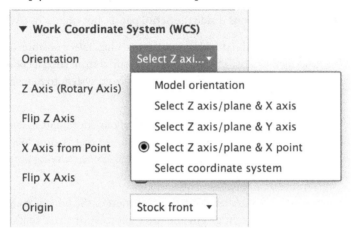

Figure 2.11: Orientation options

As you can see, there are several options we can pick in order to specify the WCS orientation:

- **Model orientation**: We get the coordinate system from our part to be machined, and we keep the same orientation of the axes (this is an option rarely used).

- **Select Z axis/plane & X axis** or **/plane & Y axis** or **/plane & X point**: We pick the rotation axis (*z* axis) first, and then we define another element to set the other two axes (*x* and *y*).

- **Select coordinate system**: With this option, we can specify any WCS inside the project to be used. This can be a WCS from our 3D model or the chuck or any submodel.

> **Note**
>
> By default, Fusion 360 considers the *z* axis as the rotation axis and the *x* axis as the radial direction that our tool is moving along. This is typical for most machines; however, yours may be different, so please check your axes orientation.

After we set the orientation of the *x*, *y*, and *z* axes, we also need to adjust the tip direction. Fusion 360 gives us the option of flipping these axes:

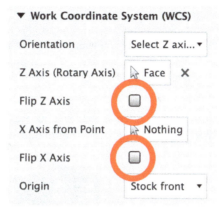

Figure 2.12: Flip axis button

Let's now suppose we click on **Flip X Axis**. What is the outcome?

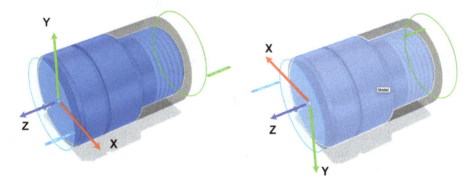

Figure 2.13: Different WCS orientation

On the left, we can see the WCS orientation before flipping, while on the right, we can see the orientation after the command is pressed. As we would have guessed, the x axis is now pointing in the opposite direction, while the z axis is unchanged. You may have also noticed that the y axis is flipped, a subtle change that leads us to our next point.

Every coordinate system in any 3D software is "right-handed." This means that if we point our right thumb along the x axis and our right forefinger along the y axis, then our right middle finger will always point in the z direction.

> **Note**
>
> If you are not able to orient all the axes the way you want after a few axis flips, the chances are high that you are trying to set a forbidden "left-handed" WCS. If this is the case, you have no choice other than to set the best right-handed WCS for your needs.

Also, it is important to remember that the z axis orientation may influence the coordinate value assumed by the tool while working. Let's check the following screenshot to understand how:

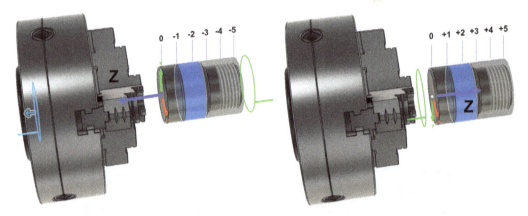

Figure 2.14: Different WCS orientation (continued)

As you can see, on the left, all the coordinates of our part are negative, while on the right, they are all positive. The same tool at the same position will be described by a different set of coordinates on the right example and on the left example.

This is also true for the resulting toolpaths; they will look the same to our eyes, but in reality, they are different and not compatible. Therefore, copying a toolpath from the right setup to the left setup will produce unexpected movements!

Also, note that the x axis is unchanged, the z axis is flipped, and due to the right-hand rule, the y axis is flipped as well.

Once we set the right orientation of our WCS, it is time to set the origin location. In order to specify the origin point, we have multiple options to choose from:

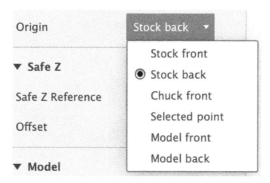

Figure 2.15: Origin point setup

Here, we can set the origin at the front of the 3D model we want to machine or at the back. We can also specify a point along the z axis where the chuck ends (known as the chuck front) or at the boundaries of our stock.

Most of the time, I choose to set the origin at **Stock front** with the z axis pointing outward because I like to use negative coordinates when machining my parts, but this is something I like to do and is not mandatory for you; you can set the origin wherever you prefer.

We should now be quite confident in all the options related to WCS placement and orientation, so let's move forward!

Safe Z

In the **Safe Z** section, we can specify where the tool can move safely in the radial direction without colliding with our part. We have two values to set, which are **Safe Z Reference** and **Offset**.

Safe Z Reference

This is a z coordinate we set as a safe position for our tool. After this z coordinate, our tool should be able to get any radial position without colliding with the stock or the chuck. Are you a bit confused?

To get a better understanding, suppose we need to move our tool toward the rotation axis (z) for whatever reason. This movement is not intended as a cutting operation, so the chuck may also be standing still:

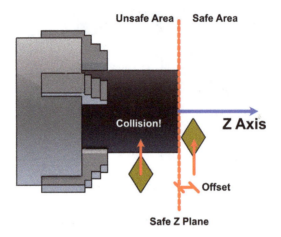

Figure 2.16: Safe Z

As you can see from the example, the WCS origin is set on the front of our stock. If we try to move our tool toward the rotation axis while located at negative *z* coordinates (on the left of the dashed plane), it will crash against the stock.

On the right of the dashed plane, we no longer risk crashing into the stock.

In this example, the dashed plane can be considered a **Safe Z** plane type—it divides *z* coordinates between safe and potentially unsafe areas.

Please note that the **Safe Z** plane refers to the tip of our tool only; it doesn't consider the tool and shank dimensions. Therefore, most of the time, we need to add an extra offset value to avoid collisions.

The typical **Safe Z** position is on the front of our stock (**Stock front**), on the opposite side of the chuck:

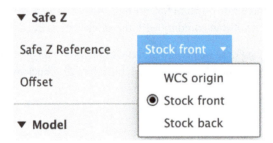

Figure 2.17: Safe Z Reference options

We can specify a different position such as **Stock back** or **WCS origin**, but 99% of the time, you will set it to **Stock front**.

Offset

As already mentioned, **Offset** is a safety margin from **Safe Z Reference** that takes into account insert and shank shape. Even if the tool shape shouldn't collide with the stock, it is still a good idea to specify a bit of offset. Can you guess why?

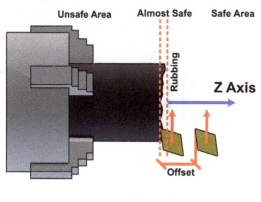

Figure 2.18: Safe Z offset

As you can see in the previous example, the insert should be able to move along the **Safe Z** plane without any collision. But in reality, the front of the stock may be cut at an angle, or it may be a bit rough and still cause rubbing (which is bad) or collisions (which is worse). Therefore, we should always add a bit of offset from the **Safe Z** plane and be sure that there is always a layer of air between the tool and the stock when not cutting.

A safety gap of at least 5-10 mm is always a good idea.

> **Note**
> Depending on the z axis orientation, a positive offset value may increase the gap or decrease it, so you should always check twice!

But thanks to Fusion 360, it is quite simple to visualize the planes when preparing our setup:

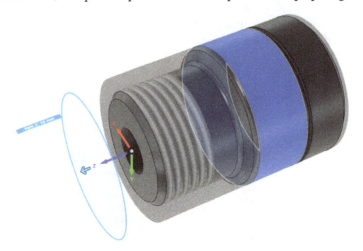

Figure 2.19: Offset plane

As you may have noticed from the screenshot (or from your screen), there is a light blue circle that we can move by dragging on the arrow that represents the **Safe Z** plane, and there is an offset of 15 mm from the stock front.

> **Note**
>
> As previously mentioned, flipping the z axis will invert the offset value. With the z axis pointing inward (with the WCS origin at the stock front), negative offset values will result in air between the stock front and our tool!

Model

In the **Model** subpanel, we can pick a model to machine among the group inside the project. Here, we can find two commands:

- **Model**
- **Spun Profile**

What are they used for?

Model

The most straightforward setup occurs when we have only a single part inside the project. However, as already discussed, we may have multiple models inside our 3D environment. For example, we may want to machine a welded assembly with multiple subcomponents, or we may have a work-holding fixture or a chuck.

With multiple parts inside the project, Fusion 360 may start panicking because it will not understand which is the model to be machined. Its default behavior is to consider all the parts available as machinable. With the **Model** command, we can pick only the model we intend to process with our lathe.

> **Note**
>
> After the selection, Fusion 360 will try to arrange stock placement and WCS again. Therefore, this should be the first command you use when beginning a setup!

Spun Profile

Spun Profile is an option that allows us to manage complex parts that have both turning features and milling features on a lathe. Most of the time, despite our lathe being such a useful machine, we may need geometries that are not axially symmetric; therefore, we cannot machine those only by turning.

For example, we may need to create a slot inside our part. Why is this a problem for Fusion 360? To answer this question, we have to know how Fusion 360 handles toolpath generation. At first, it will cut the 3D model in half by slicing it, and then it will take the cross section and use the outer contour for toolpath coordinates:

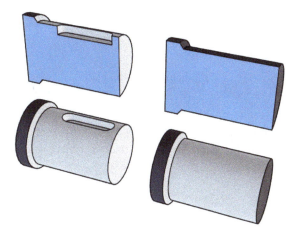

Figure 2.20: Spun Profile on/off

As you can see from the screenshot, the model on the left has a slot, and the resulting cross section is displaying it. Therefore, Fusion 360 will consider the left profile as an axially symmetric part, which, in reality, is quite different from what we want to machine. In order to avoid this type of problem, we should remove the unmachinable geometries first. As you can see, on the right part, we removed the slot and restored the cylindrical shape. This operation is called **defeaturing**.

> **Note**
>
> Defeaturing is an operation related to parts that need to be simplified. There are several possible reasons for defeaturing. For example, we may need to reduce the file dimension of a complex assembly, or we may need to simulate physics. In order to reduce computation time, we can decide to remove useless features.

Thanks to Fusion 360, we don't have to manually remove features; there is a useful option that does just that. With **Spun Profile** checked, we can select the profile to be machined by our lathe while ignoring all the features we plan to machine on our mill. We can consider this command a sort of defeaturing command.

Still confused about **Spun Profile**? Take a look at the following screenshot for a better understanding:

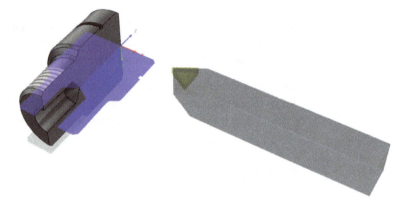

Figure 2.21: Spun Profile

As you can see, the model is not turnable since it has a heavy milled face that totally breaks its axial symmetry (it is basically cut in half). Even if Fusion 360 shouldn't recognize the cylindrical profile, after enabling **Spun Profile** in the setup, Fusion 360 still recognizes its round shape cross section (highlighted in blue in the example).

Chuck

The **Chuck** subpanel is where we specify our fancy geometry for our chuck for collision simulation. Here, we find two subcommands: **Chuck reference** and **Offset**.

Chuck reference

Chuck reference is where we set the position of our chuck. There are multiple options to choose from:

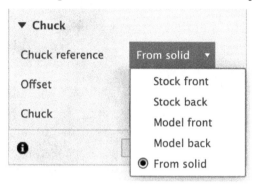

Figure 2.22: Chuck setup

As you can see, there are several options; however, most of them are self-explanatory:

- **Stock front**: We can set the chuck position on the front of the stock
- **Stock back**: The chuck will begin at the back of the stock (this is very common)
- **Model front**: This will set the chuck position on the front face of our 3D model
- **Model back**: The chuck will be positioned at the end of the 3D model
- **From solid**: Instead of specifying a plane on the z axis, we can pick, model by model, every component of the chuck if we imported one into the project

With most of the options, we specify a plane (much like we did with **Safe Z Reference**). In this case, we may call it **Danger Z Reference**; after this value on the z axis, we can be sure that our tool will collide and break something valuable.

Offset

The **Offset** value is intended as a safety margin before collision and is similar to the **Offset** value for the **Safe Z** plane (everything said for the **Safe Z** plane also applies to this value as well).

Thanks to Fusion 360, it is always clear where the **Safe Z** plane and the **Chuck** plane are located, as you can see in the following screenshot:

Figure 2.23: Chuck plane

Here, you can see that there is a green plane that we can drag along the z axis, which will be where the chuck starts. It is always a good idea to work with high offset values, at least 15-20 mm, so that we can be sure that no collision will occur.

It is important not to underestimate the strength of a tool's impact against the chuck; with several kilowatts involved, a tool fragment may become a dangerous projectile. Do not bypass machine safeties, and always wear protective glasses!

So, we have now discovered the first tab of the **Setup** command, where we explored topics such as WCS and the **Safe Z** plane; now, let's move on to the next tab.

Stock tab

Inside the **Stock** tab, we can find all the parameters related to the stock shape and dimensions. In the following screenshot, we can find a preview of the panel:

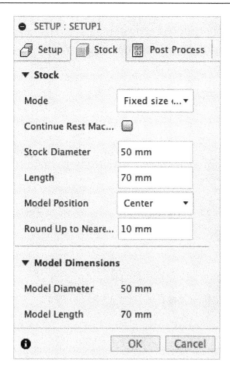

Figure 2.24: The Stock panel

Fortunately, the only parameter we really need to be looking at is the **Mode** option inside the **Stock** subpanel. It will change the way we specify the stock shape and its dimensions:

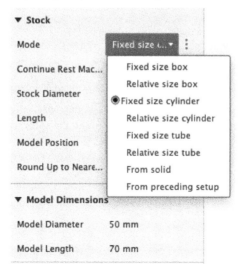

Figure 2.25: Mode options

As you can see, we can specify different shapes to machine with our lathe; some of them are for square stocks, while others are for cylindrical stocks or tube stocks. When you specify a shape, some of the options may change slightly.

So, let's check each of the shapes and options now.

Fixed size box

With **Fixed size box**, we can specify a squared stock to be machined by providing the global dimensions:

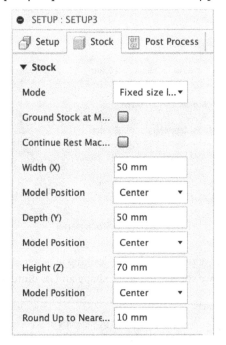

Figure 2.26: Fixed size box

As you can see, not only do we have to specify the global dimensions of the bounding box but we also need to set the **Model Position** value, which is an offset along the specified axis.

Relative size box

The **Relative size box** option is similar to the previous one since it creates a squared stock; however, it is different in the way we specify the shape. Let's check the following parameters:

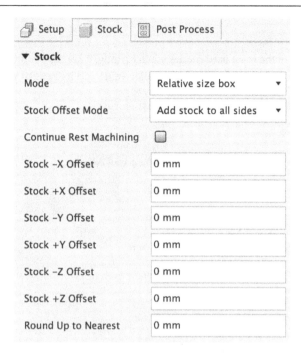

Figure 2.27: Relative size box

Here, we do not specify the bounding box dimensions for our stock, as Fusion 360 automatically calculates the needed stock size to machine our part. However, we can set extra stock thickness around our object by specifying offsets along the three axes.

> **Note**
> Why would we want to set extra stock around our object? Because our stock may not be a perfect box—it may be bent, or it may be a bit smaller, or it may be rusted. In all of these scenarios, we add extra material to be removed around the part, allowing us to be more confident about the final result.

This was the second "box mode," but to tell you the truth, it is very rare to start from a box stock when turning metals. Instead, we always start from an axially symmetric stock such as a cylinder. That's because non-round stocks will generate strong vibrations and strong impacts on our tool.

Fixed size cylinder

Fixed size cylinder is one of the most used modes when turning; it creates a round stock with specified dimensions. We just have to specify the **Stock Diameter** value and its **Length** value:

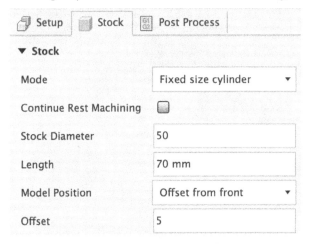

Figure 2.28: Fixed size cylinder

Similar to **Fixed size box**, we can specify an **Offset** value from the front of our stock or from the back (again, this should be used if we need extra precision regarding the final result).

Relative size cylinder

This option is similar to **Relative size box**. With this option, we don't specify a diameter and a length; we have to set how much additional stock we want around our part.

Fixed size tube

Very similar to **Fixed size cylinder**, here we create a hollow cylinder (a tube), so we need to provide the outer diameter, inner diameter, and length.

Relative size tube

At this point, you should now understand the difference between fixed size and relative size, so with **Relative size tube**, we can specify additional material around the tube needed to machine our part.

From solid

From solid is a very interesting option. Instead of specifying a default shape such as a cylinder or a tube, here, we can pick a 3D model inside the project to be the stock. This option is very useful; for example, we can use it when we need to machine a part that was created from another manufacturing process, such as a metal casting or a welded assembly.

From preceding setup

Since, as we discussed, there can be multiple setups listed one after the other, with this option, we can acquire stock dimensions from the previous setup. Please note that WCS placement has nothing to do with stock. This mode only retrieves stock dimensions. Frankly speaking, I've almost never used this option.

Finally, we completed all the stock creation modes; they should not be very difficult to master. There's one last thing to note. While looking at all the stock mode panels, the most curious of you may have noticed a recurring checkbox called **Continue Rest Machining**:

Figure 2.29: Continue Rest Machining

What does it mean? This option is very important as it allows us to keep working on the stock machined on a previous setup! For example, imagine that we started machining with our lathe on the first setup. After turning the part, if we need to keep milling our stock, we need to create an additional setup. However, with **Continue Rest Machining** checked, Fusion 360 will consider the already removed material and will not try to machine air (where the material was already removed from previous operations)!

We have now completed the second tab of the **Setup** command, where we discovered many options for stock placement and shape. There is just one more tab to discover, and that is related to postprocessing.

The Post Process tab

There is just one more tab to analyze before we can say that we have mastered setup options: the **Post Process** tab.

But first of all, what is **postprocessing**? It is the junction between our CAM module and our CNC machine—a sort of translation between our 3D model language and our machine language.

The language spoken by our 3D software is quite advanced, made up of complex 3D shapes, materials, and colors. However, even the most expensive CNC machine—whether it is a lathe, a mill, or a 3D printer—is quite dumb; it is mainly able to understand simple numbers. The **postprocessor**, a subprogram that manages the postprocess, converts the most complex information from our CAM into simple formulas and coordinates that our machine can understand. The language created by the postprocessor and understood by the machine is commonly called **G-code**.

> **Note**
>
> G-code is a proper programming language. We could write very simple programs ourselves, but we would need to follow specific rules and a strict syntax; any error, or even a misplaced comma, would cause a machine emergency.
>
> Also, every CNC machine may speak a slightly different G-code language, so a program that works flawlessly on our CNC may be unreadable by another. Due to this, we may need to use different postprocessors according to the machine we want to use.
>
> Fusion 360 comes with a decent set of postprocessors that work for several common machines, but there's a chance yours may not be compatible. Before buying Fusion 360 or a CNC machine, you should always check that they can work together by default or by a third-party plugin. If not, don't waste your money!

This is not the best starting point to understand G-code and postprocessing, since we are still at the beginning of our journey inside the CAM world. So, for now, let's just remember the key concept behind it: it is a translation process.

In the following screenshot, we can find the **Post Process** panel inside the **Setup** command:

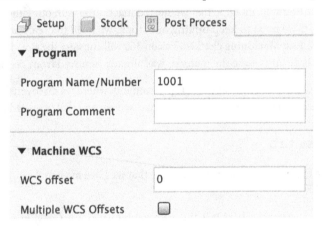

Figure 2.30: The Post Process tab

As you can see from the screenshot, this panel is quite slim, but there are some important options to understand.

Program Name/Number

As we mentioned, G-code is a set of instructions written in the language spoken by our CNC machine. A typical set of instructions to machine a part may consist of hundreds of instructions and commands; therefore, the instructions have to be stored in a file.

With the **Program Name/Number** option, we provide the name for this file once it has been created (do keep in mind that certain machines may want a custom program name convention with a strict syntax).

Program Comment

Since a G-code file is written in a language easily understandable only by a machine, if we decided to open it with our text editor, we would almost certainly lose the idea of what the set of instructions is used for. That's because when G-code is opened with a text editor, it consists of thousands of coordinates listed.

Using **Program Comment**, we can add a line to the beginning of the file that will help us to retrieve important information in the future. This comment is never read by the machine; therefore, we can write it in human language.

As an example, we may want to add a few words related to stock dimensions or WCS placement or the material to be machined, or anything that may help us in the future.

Machine WCS

This is the last section of our setup, but unfortunately, at the moment, it is way too advanced for us to understand and is slightly beyond the scope of this book. Saying that, we will return to this topic in *Chapter 8,* when we will set up a multi-placement milling operation; it is mostly used when we need to change the machine coordinate system—for example, with multiple stocks to be machined.

For now, you can leave the **Machine WCS** settings as is.

And with that, we have completed quite a deep overview of all the possibilities available inside the **Setup** command for turning operations. What a milestone!

Summary

Congratulations—that's the end of the second chapter! I hope that you are getting more confident with the first steps inside the Fusion 360 CAM module.

Let's quickly recap what we learned during this chapter. We started by discovering where we can find useful example projects that we can use in our setups, before exploring the **MANUFACTURE** module interface.

After that, we learned how to import a chuck (or any other generic workholding stuff inside our project) and reviewed all the options available inside the **Setup** panels. Finally, we got a hint about postprocessing and G-code.

Please always remember that a good setup is the best way to approach better machining without risking unwanted collisions or a bad WCS placement.

This chapter has taken us one step closer to spinning our chuck. In the next chapter, we will discover where to find the tools to machine our part.

3

Discovering the Tool Library and Custom Tools

In this chapter, we are going to discover the built-in tool library of Fusion 360 and how to import new tools to use in our machining strategies. However, unfortunately, Fusion 360 doesn't come with many turning tools inside the default library.

Are we forced to only use these tools? Of course not!

We have several options to expand our toolset, all of which we will explore in this chapter. This includes creating a tool from scratch, an option that is quite time-consuming but allows for the highest flexibility.

Another option we will explore is importing an entirely new library alongside the default tool library. This will give us the tools we need but also many more tools we don't. The other option we will explore is using a third-party plugin.

The goal of this chapter is to get you used to multiple new environments where you can find useful resources about tools and their geometry.

Do not underestimate the importance of a proper tool library – using the right tool for the right operation is critical as doing so is the key to long-lasting tools, good surface finish, and good tolerance!

So, in this chapter, we will cover the following topics:

- Discovering the tool library
- Creating a new tool
- Importing a third-party tool library
- Importing new tools with a plugin (CoroPlus)

Technical requirements

To follow along with this chapter, you need to have read most of the previous chapters; the first chapter is mandatory, but the second chapter can be skipped as we won't mention much of what was stated there.

You will also need a working license for Fusion 360 and an internet connection to download additional content.

Also, note that at the end of this chapter, we will mention a third-party plugin called CoroPlus that requires its own paid license. Though it is a bonus, it is not strictly necessary for our CAM journey.

Discovering the tool library

Before starting any turning operations, we must take a look at the tools that Fusion 360 has to offer. Unfortunately, it doesn't come with a lot of tools to pick from; the options are quite limited, to say the least.

Where do we find the tools? In the **Tool Library**, of course:

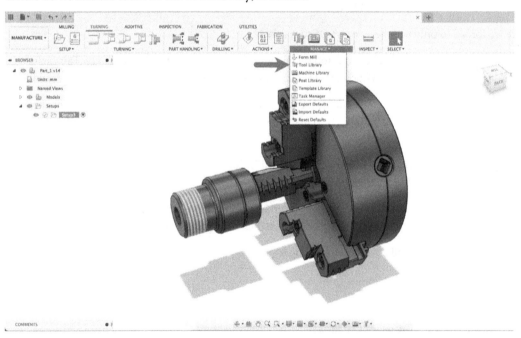

Figure 3.1: Tool Library location

Let's click on the icon and open the library:

Figure 3.2: Tool library

Inside the tool library, we can find many different tools; these are not only related to turning – there are also milling tools, tool holders, laser nozzles, and so on.

Most of the time, the large number of tools available may be a bit overwhelming. To find the tools that we need, we have to know how to navigate inside the tool library. Let's break down the previous screenshot:

1. In the left column, there is a list of all the available libraries. To search for all the built-in tools, we must select **Fusion 360 Library**.

2. To reduce how many tools and holders are available to us, we can set a filter on the right-hand column; as you can see, we have just set a **Turning** filter, so only turning tools are listed.

3. At the moment, the number of tools available is quite limited (this is especially true if we remember the variety of operations and shapes that we discovered in *Chapter 1*).

Don't worry – even if there aren't many tools built in, Fusion 360 still allows us to expand the tool library through other methods. As mentioned in the introduction, we can create tools ourselves, import third-party libraries, or use plugins. In the following pages, we will discover these different options.

Creating a new tool

The first way to add tools to the limited built-in tool library is to create our own! This is the least expensive option but is also a bit more time-consuming.

Fusion 360 allows us to create a new tool from scratch and set all the needed parameters. Of course, we do not have to reinvent the wheel; we can simply start from the real tools we have at our disposal and copy their geometry. Alternatively, we can search for the tool we need inside a turning tool brand, copy it, and then buy it if it fits our needs.

Imagine that we need to add a new tool, such as a new finishing tool. We do not have to create a tool with our imagination; the best idea is to start looking on tool brand websites for their finishing tools lineup or check what we already have at our disposal on the shelf.

Getting a sample tool

Now, let's suppose we found an interesting tool on our favorite tool shop website with the following specifications:

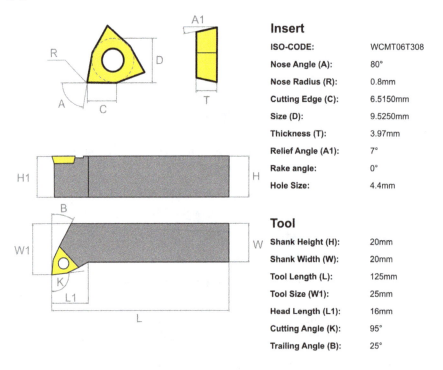

Figure 3.3: Example tool specifications

If we use quality tools, we should always find all the pieces of information needed to set up our tool. As shown in the preceding figure, we are provided with all the details we learned about in *Chapter 1*; there are just a couple of things I'd like to point out since we haven't found those yet.

One useful thing to know is that almost every tool has a specific **ISO-code** assigned to it. For example, the insert we are about to use is completely described with the ISO-code only (**WCMT06T308**).

Therefore, we can refer to such an insert by specifying all the geometry parameters or by simply using its code. Please note that standard ISO-codes are not brand dependent, so we can find the same insert made by different brands.

You may have noticed another angle called **Trailing Angle**, which is the angle that allows our shank not to impact the stock while contouring; the higher the angle, the better the accessibility.

Now that we have all the specifications of our tool, we can create a clone inside Fusion 360.

Creating a new tool

To create a new tool, the first step is to open the tool library:

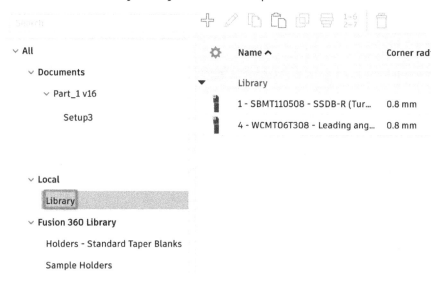

Figure 3.4: Local tool library

As we've already discovered, there is the built-in **Fusion 360 Library**, which contains a lot of tools. However, note that this library is read-only, which means that we can access all the tools, but we cannot modify them or add more to the list.

Instead, we can use our **Local library** to store all the newly created tools. Once selected, we will find all our tools listed. If we want to add one, we can just press the + button (if the button is unavailable, you are probably inside a read-only library!).

> **Note**
>
> Please be careful not to save tools inside the current document. Saving tools inside the document will make them inaccessible when working with any other document.

Let's click on the + button. We will be presented with a list of all the possible tools we can create. For this, we are looking for an external turning tool, so click on **Turning general**:

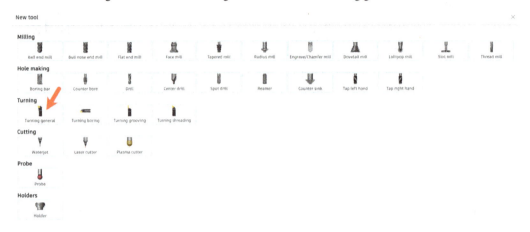

Figure 3.5: New tool type

Now, we will be presented with a new window with multiple tabs – **General**, **Insert**, **Holder**, **Setup**, **Cutting data**, and **Post processor** – where we can specify all the necessary data. We will go through each tab now. Don't worry – we don't have to fill in every detail available; there are a few shortcuts!

General

On the **General** tab, we can add the main information related to the tool we are about to create:

Figure 3.6: The General tab

Inside the **Description** field, we can add any note related to the tool's intended usage, surface coating, price, and so on.

With **Vendor**, we can specify the supplier, and with **Product id**, we can specify the vendor code for future purchases – these are very handy fields to have.

Finally, we can add a **Product link** to the seller's web page to make things even easier.

Let's now move to the following tab called **Insert**.

Insert

On the **Insert** tab, we can specify all the specifications for our cutting insert. As shown in the following screenshot, there are many parameters. We could waste a lot of time manually typing all the values one by one, but most of the time we can simply copy and paste the needed ISO code inside the text box:

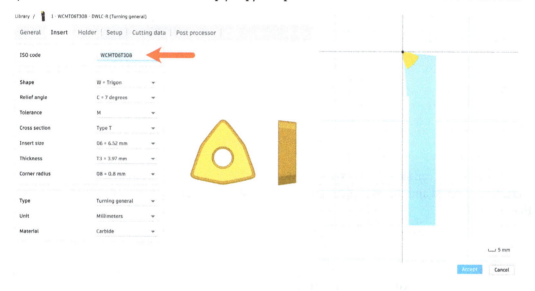

Figure 3.7: The Insert tab

Here, I copied the ISO code from the previous insert datasheet, and the insert geometry was automatically created with the same shape. However, the shank still looks quite different from what we saw previously; we will discover how to change it in the following tab.

Holder

Unfortunately, Fusion 360 doesn't support any shank coding, so we have to edit shank parameters manually. The following screenshot shows the **Holder** tab, where we can create the desired shape:

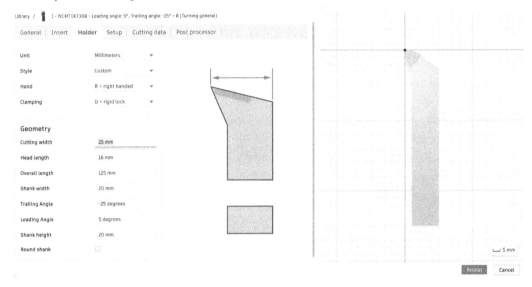

Figure 3.8: The Holder tab

As you can see, we can simply set **Style** to **Custom** to be able to set all the geometry specifications. Here, I have copied and pasted all the values from the shank; the 3D preview is now quite close to the real tool!

> **Note**
>
> You don't always have to specify the shank geometry; if you are sure that no collisions are going to occur, you only need to use the insert geometry!

Setup

We do not have so much to modify on the **Setup** tab:

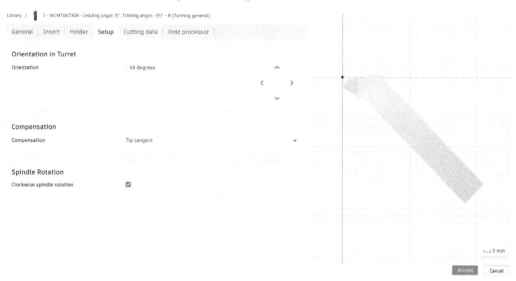

Figure 3.9: The Setup tab

Here, we can set the shank rotation, also known as **Orientation**, but most of the time, we use 0° for external operations and 90° for internal operations.

We can also change the **Compensation** parameter, which allows us to shift the cutting origin of the tool (more on this in *Chapter 10*). However, for the sake of simplicity, let's leave it set to **Tip tangent** (do not change it unless you know what you are doing!).

Cutting data

Cutting data is one of the most interesting tabs. Here, we can specify the default cutting parameters to use with this tool. However, it is not mandatory to insert these values – we can manually specify them later since, as we already know, parameters greatly change on different operations. We can set default cutting parameters if we already know that most of our machining with this tool will have similar geometries with similar cutting parameters.

Now, let's review the main options we can set:

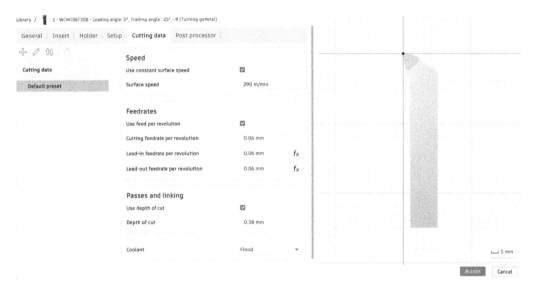

Figure 3.10: The Cutting data tab

If our lathe can smoothly control rotation speed via an inverter, we can tick the **Use constant rotation speed** option. With this option enabled, our lathe will automatically tweak the rotation speed to achieve a constant cutting speed. As already mentioned in *Chapter 1*, the chuck will rotate faster when cutting on a smaller diameter and will spin slower when cutting on a bigger diameter. Also, you can use the **Surface speed** option to set the default cutting speed for the tool.

Inside **Feedrates**, we can specify the default cutting feed for the tool. Two values to note are **Lead-in feedrate** and **Lead-out feedrate**; with these values, we can specify different feeds before engaging and after leaving the stock, respectively. Most of the time, unless we are looking at maximum productivity, we can use the same feed value for them all.

Under **Passes and linking**, you can see an option called **Depth of cut**. If you want to calculate the cutting parameters, please always refer to the formulas shown in *Chapter 1* (however, it is always a good practice to take a safety margin of at least 20-30% over the results of the formulas).

The last parameter to specify is the **Coolant** method, which mostly depends on the machine and the material we want to cut. The default cooling method is **Flood**. With this type of cooling, our part is constantly sprayed with a lubricant coolant. This type of coolant is very effective since it allows us to drop cutting temperatures and enhance cutting performance.

As a rule of thumb, always cool your parts when machining. Cooling your part too much will never be a problem unless the material can absorb it (for example, wood)!

Post processor

The last tab to mention is the **Post processor** tab. Inside this tab, we can help improve the post-processor results with a few features (if our machine supports them):

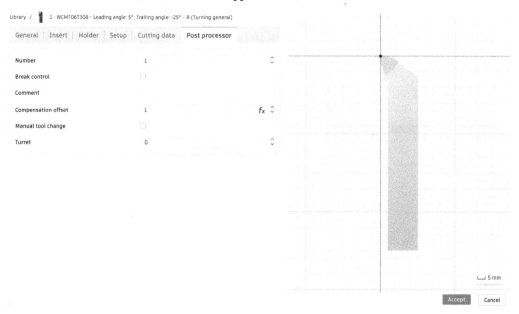

Figure 3.11: The Post processor tab

If our lathe can change tools automatically, we can specify where the tool is located inside the housing using **Number**. As shown in the preceding screenshot, at the moment, it has been set to be at position number 1. This means that our lathe, when changing tools automatically, will expect to find this tool inside housing number 1. We should always check the correct location inside the housing.

Another feature we can specify (if our lathe can manage it) is **Break control**. If we tick this option, before going forward with a new machining operation, our lathe will check whether the tool is in a good shape or whether it is broken.

If our lathe doesn't have an automatic tool change option, we should tick **Manual tool change** and skip all the other parameters on this page.

And with that, we have analyzed most of the options available when creating a new tool inside Fusion 360. However, for the lazy users among us, there are faster and more painless ways to get new tools: using third-party options.

Importing a third-party tool library

Creating tools is nice and cheap, but it can be a bit time-consuming if we need several new tools. Another option is to increase the built-in tool libraries by downloading third-party libraries and importing them inside Fusion 360.

So, how can we do that? First of all, we should enable cloud libraries. You can find the **Enable Cloud Libraries** option in the **General** preferences area, under **Manufacture**. Once found, be sure to tick the box:

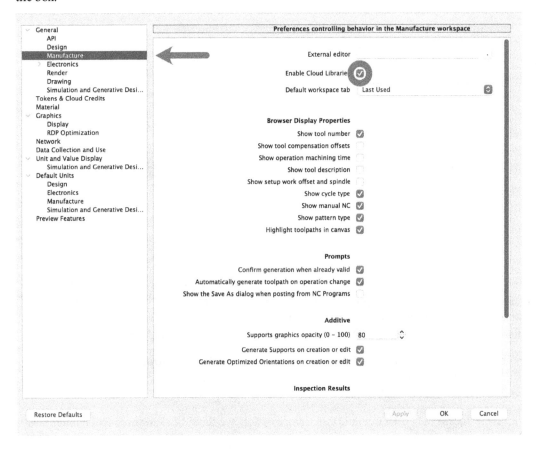

Figure 3.12: Manufacture preferences

This option will create a bridge between our tool library and our projects.

After this, we can download a new tool library. You can find many libraries online; some of them require a subscription, and some just an account. For the sake of simplicity, we are going to use a sample tool library by nyccnc.com.

Once you have downloaded the attached file, unzip it. There should be a .json file inside the archive. To import it into Fusion 360 libraries, we have to import it into the **Assets** folder. This folder is located in the side panel of the project root:

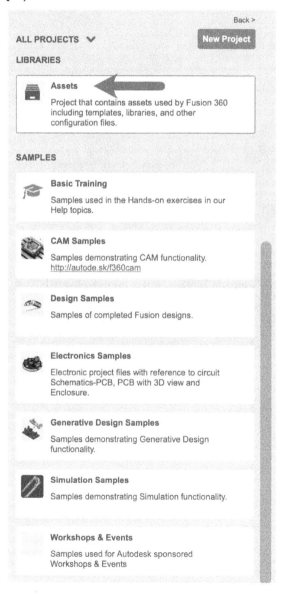

Figure 3.13: The Assets folder

Once inside the **Assets** folder, open the **CAMTools** subfolder (*1*):

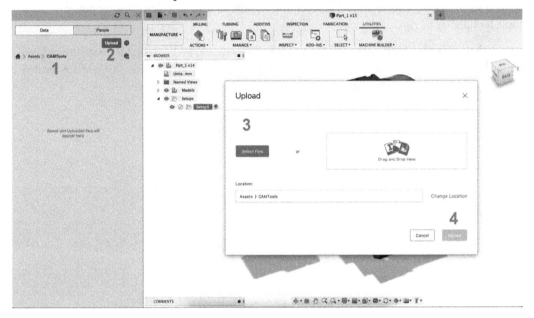

Figure 3.14: Import library

With **CAMTools** open, click **Upload** (*2*). Then, select the downloaded library (*3*) and click **Upload** to upload the file (*4*).

And that's it. If we open the tool library again, we should find a few differences from before:

Figure 3.15: Finding the new tools

First of all, we should have noticed a new **Cloud** section (*1*), under which we will find our new third-party library (*2*). You should also see several new tools listed (*3*); if we click on an element, we can find the tool preview (*4*) and the tool specifications (*5*) too.

At this point, we could move on to the next chapter since we already know how to add lots of tools to our library, but we would miss one very useful option: using a third-party plugin. This third way is definitely worth mentioning.

Importing new tools with a plugin (CoroPlus)

As you may expect, the limited turning tool library of Fusion 360 is a problem that we don't have to face alone. As already discussed, several tool companies have created tool libraries for Fusion 360, but Sandvik Coromant has taken things to the next level, creating a plugin that has so much more to offer.

First of all, what is a plugin? A plugin is an additional piece of software that gives the main software more features and commands. There is a plugin browser inside Fusion 360 – consider it like an app store – where we can look for new features to add.

The plugin we are about to discover is called CoroPlus, and we should consider it more like an advanced CAM module rather than a mere tool database. Not only do we have access to thousands of tools, but we can also use it to calculate cutting parameters and so much more.

There are only two main drawbacks: it is licensed under a subscription and it has limited support for turning operations. However, Sandvik allows us to try the software for 30 days free of charge. It is a very good plugin and I think it is worth the cost, but if you still feel disoriented with turning, I invite you to give it a try later.

If you are not scared yet, let's review how to install this plugin and give it a try! First, you need to find the plugin browser; make sure that you are in the **UTILITIES** tab (*1*). Then, open the **ADD-INS** subpanel, and launch **Fusion 360 App Store** (*2*):

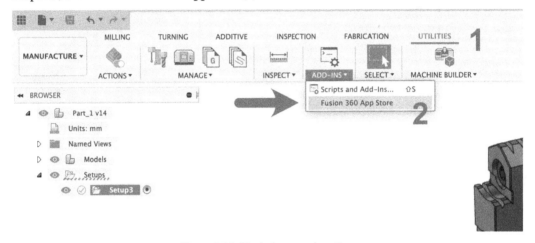

Figure 3.16: Plugin browser location

We can easily find the CoroPlus plugin by filtering for CAM plugins or by typing `CoroPlus` in the search bar:

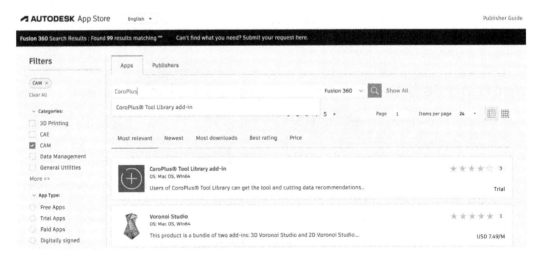

Figure 3.17: CoroPlus plugin

Once located, select the plugin and download it, making sure to choose the platform option compatible with your setup:

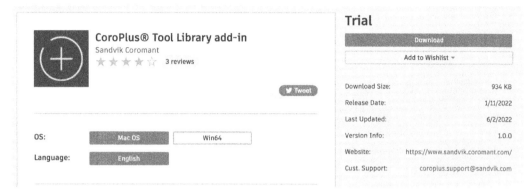

Figure 3.18: Plugin download

The installation process should be very straightforward – we only need to log in with our Autodesk account and follow the instructions on the screen!

So, the installation went smoothly, but how can we open our plugin? You should be able to find the plugin inside the **UTILITIES** tab, like so:

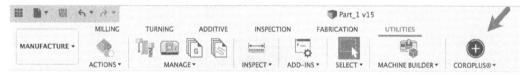

Figure 3.19: CoroPlus icon

However, if you can't see it in the tab, you may need to load the plugin first. To do that, go back to the **ADD-INS** menu and select the **Scripts and Add-Ins…** option:

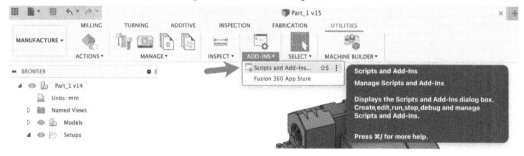

Figure 3.20: Add-ins startup

This command will open the following panel, which allows us to manage all the extensions and plugins (for example, if a plugin causes our project to keep crashing, we may disable it):

Figure 3.21: Add-ins management

To activate the new plugin, make sure that you are on the **Add-Ins** tab, select the plugin's name, and then hit **Run**. Please note that a good idea is to tick the **Run on Startup** option to automatically load the plugin at startup.

We are now ready to launch the plugin and log in to our Sandvik account:

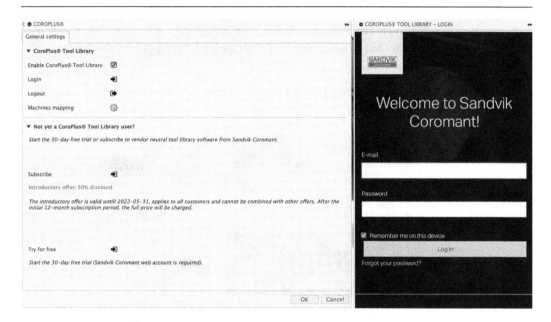

Figure 3.22: CoroPlus login panel

Once we've logged in, we can unleash all its potential. Grasping all of the features that this plugin can offer is way beyond the scope of this book, so for now, we'll just show you how to gain access to the tool library.

If we want to use a new tool, we have to create it from the Sandvik online database, save it to our local CoroPlus database, and then send it to the Fusion 360 library. It is quite a tricky process at first, so let's find out how to do it.

As shown in the following screenshot, the first time we open the tool selection window, the interface is pretty much empty:

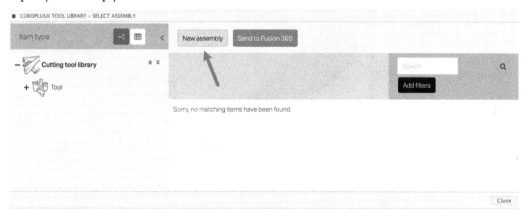

Figure 3.23: CoroPlus local database

There are no tools to select here because we never downloaded any from the online database. Therefore, the first thing we need to do is create a new assembly for our tool. It is called an assembly because the insert and shank are combined together.

You should now be presented with the following screen, where we can search for our tool using the **Item type** tree or by using **Add filters**:

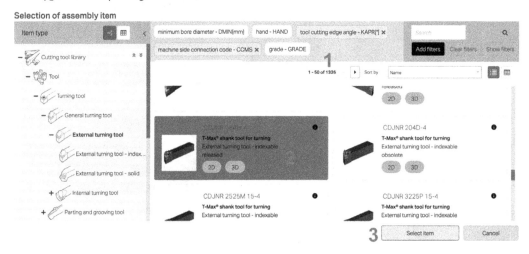

Figure 3.24: CoroPlus tool selection

As you can see, we have so many tools to search from, even after we have activated a few filters (*1*). Now, let's suppose we have finally found the tool we need – we just have to select it (*2*) and click on **Select item** (*3*). We should now be presented with the following screen:

Figure 3.25: CoroPlus tool edit

Here, we can specify the tool's name for future reference inside Fusion 360 and check all the details of the selected tool. Another nice feature to have is the 3D preview of the shank with the insert mounted; sometimes, we may get confused with all the geometry parameters, but I think that having a 3D preview of the tool can help us visually check whether we picked the tool we need.

Once we are sure that the tool is correct, we can click on **Save** to get back to the CoroPlus local tool library, which should now no longer be empty:

Figure 3.26: CoroPlus tool select

As a final step, we just have to select our tool and click **Send to Fusion 360**. Easy enough, isn't it?

Please note that at the time of writing, CoroPlus is still not fully compatible with Fusion 360 turning operations – it works great for milling but the turning part is still a work in progress. You should think twice before buying it if your main target is a lathe.

However, as we will find out in *Chapter 4* and *Chapter 5*, we can still use CoroPlus to calculate turning parameters, which is probably its best-selling feature.

After all that, we now understand how and where to find all the possible different tools that we may ever need for machining.

Summary

Congratulations! We have come to the end of this chapter. I hope that you feel a bit more confident about where and how to look for new tools to use.

Let's quickly recap what we learned in this chapter. First, we discovered the built-in tool library and how to filter the tools to find what we needed. After, we learned three different ways to add more tools that we may need to machine our part, including how to create a new tool from scratch and how to import third-party libraries and plugins.

This is a beginner-level book, so there is not time enough to increase the level of detail on this subject; instead, I encourage you to dig more into important tool brand websites, where you will find a lot more that we were forced to leave untouched!

If you want to start spinning the chuck, please move on to the next chapter, where we are going to implement turning strategies to machine our first part!

4

Implementing Our First Turning Operation

In this chapter, we will practice our first basic facing operation. Through doing this, we will discover how to get all the cutting parameters using the Sandvik CoroPlus plugin and how to import them inside Fusion 360. Plus, we will deeply analyze all the options available in this first **Facing** command since most of them will occur on other types of machining as well.

After this, we will check how to simulate tool movements to check for collisions or other errors, and finally, we will take a fast look at how to generate G-code.

The goal of this chapter is to lay concrete bases for future setups. As we will discover, the same key concepts and commands are encountered over and over again throughout the **MANUFACTURE** environment; understanding the main commands and options for a basic feature will give us a solid ground for further developments.

In this chapter, we will cover the following topics:

- Setting up a facing operation
- Discovering tool simulation
- Checking the generated G-code

Technical requirements

In this chapter, we are going to start from an existing setup (like the one created in *Chapter 2*), therefore you are supposed to have a stock and a WCS already set.

We are also going to use Sandvik CoroPlus to evaluate the cutting parameters faster without calculating all the values by hand; therefore, having a CoroPlus license might be a good idea. However, at the moment, you should already be able to roughly estimate all the values on your own.

Setting up a facing operation

One of the first operations we always have to implement is part facing. You may be wondering why. The reason is quite simple: we should always first perform those operations that will give us a true reference point for further operations or for measuring. Since, most of the time, the origin is on the front of the stock or the front of our part, a first facing operation will allow us to measure where the origin is located properly.

As you can see here, the origin is set to be on the front face of our stock:

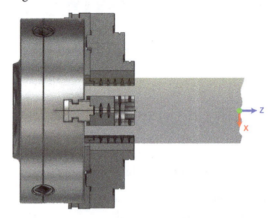

Figure 4.1: Lack of reference on the stock face

However, since the stock is never precisely cut, we may have a bit of trouble trying to take measurements from the origin!

Let's now suppose that we need to remove 3 **millimeters (mm)** of material in front of the part from a low-alloy steel stock with a diameter of 55 mm:

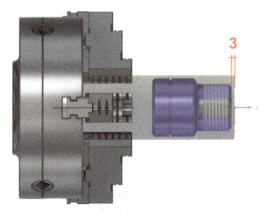

Figure 4.2: Material to be removed

At the moment, we will focus on facing the front side of our stock, while in *Chapter 5*, we will take a look at the other turning operations to machine the rest of the part.

Now that we understand the operation that we have to implement first, we should take all of the formulas and use a calculator to find the cutting parameters we must use. Luckily there is a workaround that will save us lots of equations: if you are a bit lazy like me, we can "cheat" a little and use CoroPlus instead.

As already mentioned in *Chapter 3*, CoroPlus not only gives us access to a wide range of tools but can also help us with parameter evaluation. In the next pages, we will find how to use CoroPlus at this scope.

Using CoroPlus to calculate cutting parameters

Let's just open CoroPlus and set our first facing operation step by step. Having opened CoroPlus, we should be presented with the following selection screen:

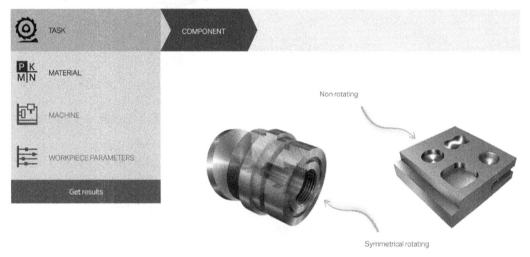

Figure 4.3: Operation type

As you can see, the interface is simple, and it helps us in the selection process. It is quite intuitive since it graphically shows the operations we can set up. Inside the first selection window (**COMPONENT**) we can choose between **Symmetrical rotating** (turning) and **Non rotating** (milling). Let's click on **Symmetrical rotating**.

We should now be taken to a second selection window that is specific for turning operations only. Of course, the more we move forward with selections, the more the operation-type range will narrow until there is only one left:

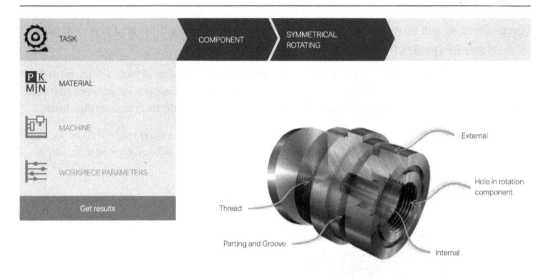

Figure 4.4: Turning operation

As you can see from the screenshot, this second selection window is called **SYMMETRICAL ROTATING**, and here, we find turning operations only. We can still choose between a great variety of turning operations (we already described most of these in detail in the first chapter, so we will skip a recap and focus on our facing example).

Facing is an external operation; therefore, we have to click on **External**. This will lead us to a third and final selection window, **EXTERNAL**, where we can choose between all the external operations:

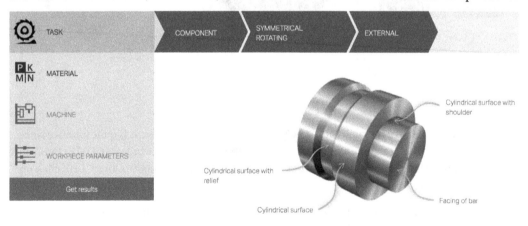

Figure 4.5: External turning operations

We should have already spotted what we need; just click on **Facing of bar** to quit the selection interface and start with the parameters calculator.

This will allow us to set all the boundary conditions and will calculate the cutting parameters we need in just a few clicks:

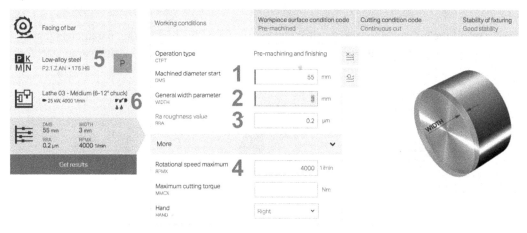

Figure 4.6: Boundary conditions

As you can see, we can specify all the parameters found in the cutting power formula. Please note that almost no parameter is mandatory, but the more we specify, the better the result; let's look at the most important ones:

1. **Machined diameter start**: This is the outer diameter of the part we need to face; in our case, the stock is **55** mm.

2. **General width parameter**: This is the distance between the front of the stock and the front of the part; in our case, it is **3** mm.

3. R_a **roughness value**: This is the desired roughness; smaller values will result in smoother and shinier surfaces, while bigger values will result in rougher surfaces. If we suppose that our example face has to be smooth and shiny, then we should set roughness to **0.8** μm (microns); this is quite a small roughness. Choosing between a higher or a lower roughness depends on the usage of the machined part; we should never target a shiny face if not really needed because a smaller roughness doesn't come for free—the part will be more expensive and longer to machine.

4. **Rotational speed maximum**: This value is the maximum RPMs achievable by our lathe. It is not mandatory to specify but it is still an important parameter to set since it will influence our cutting speed. I plan to machine the example part on a **HAAS ST-25** lathe model. Its maximum rotation speed is **4000** RPM; therefore, I should not exceed this limit on the settings.

5. **Stock material**: Here, we can specify the material we want to machine. Please note that different materials such as steel or aluminum will have totally different cutting values, so be sure that you don't skip this setting. In our case, as already mentioned, the example part will be machined out of a low-alloy steel stock.

6. **Machine information**: Here, we can set the size and the power of our lathe so that CoroPlus won't exceed its limits while calculating cutting parameters. In the example, my machine has a cutting power of around **25 kilowatts (kW)**; therefore, I set the power accordingly.

After we set all the information, we can click **Get results** to get the best cutting parameters that fit our current scenario. In the following screenshot, we can find the suggested cutting parameters for our facing:

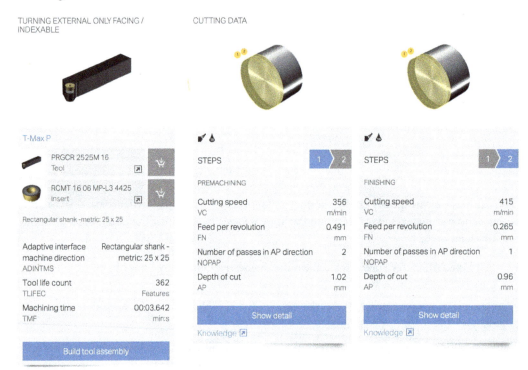

Figure 4.7: CoroPlus results

This screenshot is quite dense with useful information, but let's review the most interesting parts:

1. **Suggested tool** and **Suggested shank**: CoroPlus will evaluate which is the best machining solution according to productivity and costs. If we don't like the first proposed result, we can move to the next, even if it may be less optimized. As you can see, CoroPlus recommends that we use a round insert both for roughing and for finishing. The suggested ISO code is **RCMT1606M**.

2. **Tool life count**: This is the number of machined parts we can obtain with a single insert. As you can see, at the current cutting parameters, our tool will last for **362** parts before wearing too much or breaking. This is an important value we can use for cost evaluation.

3. **Machining time**: The time spent on the facing operations. The estimated duration for the machining is around 3 seconds. This piece of information is very handy for cost evaluation as well.

4. **PREMACHINING**: Here, we can find all the parameters of our first roughing passes. As you can see, CoroPlus is suggesting we perform two passes at a cutting depth of 1.02 mm and a feed step of 0.491 **mm per revolution (mm/rev)**. The suggested cutting speed is **356 meters per minute (m/min)**.

5. **FINISHING**: After the first roughing passes, CoroPlus is suggesting we perform one final finishing pass to get the right surface quality. As you can see, this final cutting pass is performed using different parameters that will cut slower but will result in a better surface finish. For this pass, the cutting feed is set to **0.265 mm/rev**, and the cutting depth is now set to 0.96 mm, while the suggested cutting speed also changed to **415 m/min**.

I hope that you now better understand the potential of this plugin; as we discovered, with a few clicks, not only could we define every piece of information needed to set up our operation but we also got important data to estimate the final part cost.

It is now time to take all these results and bring them into Fusion 360.

Entering the cutting parameters into Fusion 360

Getting all the required parameters for our first facing was not that hard, was it? Now, all we have to do is copy and paste all the provided parameters into Fusion 360.

First of all, where do we find the **Facing** command? If you cannot find it, be sure to be inside the **MANUFACTURE** module (*1*), look under the **TURNING** tab (*2*), and there it is:

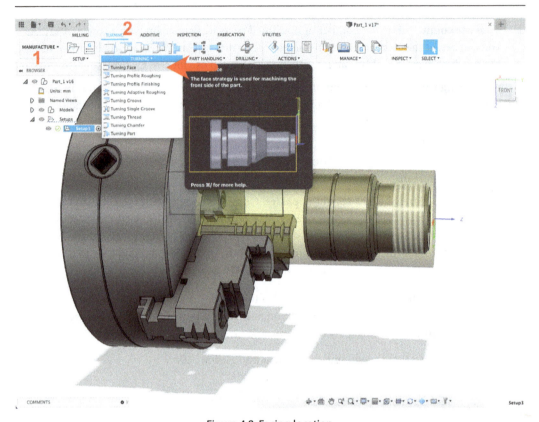

Figure 4.8: Facing location

Inside this panel, we can also find all the other turning strategies:

Figure 4.9: Turning operations drop-down menu

We will explore all of these in *Chapter 5*, but for now, we are focused on facing, so please click on **Turning Face** to open the **FACE** panel where we can specify all the settings. There are several tabs to look at, so let's go through them one by one.

Tool tab

The **Tool** tab is one of the most important tabs for any CAM operation. Here is where we have to set the tool to use and some of its cutting parameters. There are many options to manage:

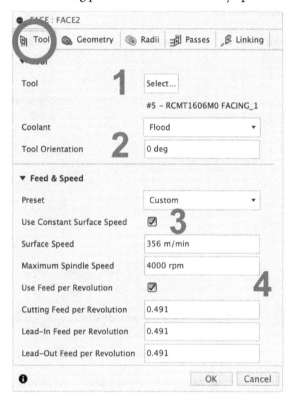

Figure 4.10: Facing's Tool tab

As you can imagine, we already have all the information needed to fulfill this tab, so just copy and paste the information from the CoroPlus results; let's look at some of the important options:

1.   **Tool**: Here, we set the tool we want to use. In the previous pages, we found that CoroPlus suggested we use an insert with the following ISO code: **RCMT1606M0**. As we discovered in *Chapter 3*, there are several ways to create an insert from scratch or to import it from external sources; therefore, we will skip tool creation.

2. **Tool Orientation**: As with most external turning operations, we leave it as **0 deg**.

3. **Use Constant Surface Speed**: With this option (if supported), we can let the lathe adjust RPMs automatically while facing. As we should remember from what was found in *Chapter 1*, at a constant chuck rotation, the cutting speed will drop closer to the rotation axis. With this option, our chuck will spin faster when cutting closer to the center.

4. **Cutting Feed per Revolution**: Here, we simply paste the resulting value from CoroPlus, which is **0.491**. Please note that we don't have a set cutting depth yet; however, we will do that when we look at the **Passes** tab shortly.

We now set everything needed for our tool, so we can move to the next tab: **Geometry**.

Geometry tab

Inside the **Geometry** tab, there are commands to set the target plane to machine on our facing:

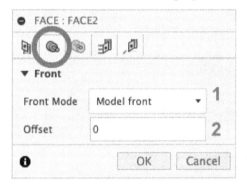

Figure 4.11: Facing's Geometry tab

Since facing is a relatively simple operation, there isn't much we have to set on this tab:

1. **Front Mode**: With this option, we can decide which plane we want to face. In our example, we need to remove all the stock from the front of our part, so we will just leave the default value as **Model front**; this means that our tool will remove material until it touches the front face of our part. If we had to face other planes, we may decide to choose **Selection** from the same drop-down menu instead.

2. **Offset**: This is a value to shift the **Front** plane along the z axis. For example, if we want to leave a certain thickness of material above our model face, we can add a positive value. As you can see from the following screenshot, after setting a value of **1.5** mm as the offset, there is a bit of stock left on the front of our part:

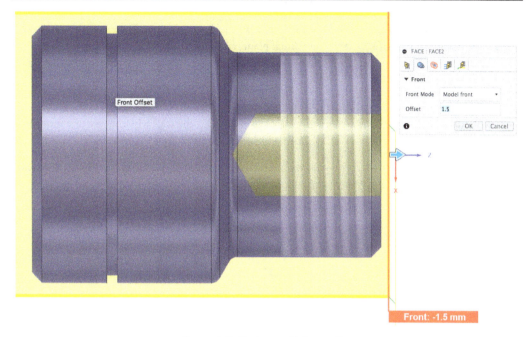

Figure 4.12: Facing tool's front offset

> **Note**
>
> Why does our label display an offset of **-1.5 mm** while on the command we wrote **1.5** mm?
> Very simple: the origin is in the front of our stock, so any front offset value will be negative!

Since we plan to set up one roughing operation with two passes of 1.02 mm and one finishing
pass of 0.96 mm, we may set an offset of **0.96** mm on the first facing and an offset of **0** on the
second facing. In such a scenario, we have to create two distinct facing operations one after
the other; however, there are other ways to leave stock around our part for future passes, so
let's set the offset back to **0** for now.

This was all you had to know about the **Geometry** tab. Let's now review the third tab related to radial
coordinates, called **Radii**.

Radii tab

This tab is very easy to understand but also very important since it manages the safety positions for
our tool; we will find it on any CAM operation.

> **Note**
>
> With milling operations, this tab will be called **Heights**, but the idea behind it and its usage is
> the same as what we are about to discuss.

Up until now, we already defined notable coordinates along the *z* axis only. For example, you may recall that we discussed the **Safe Z** plane, the **Chuck** plane, and the **Front** plane; those are coordinates on the *z* axis that specify where our tool is likely to collide with the stock or with the chuck (all those planes were set in *Chapter 2*).

With the **Radii** panel, we can specify notable planes on the radial coordinate instead. Those planes can control, for example, where the tool will retract or how deep the tool will cut. We are about to check all those planes, so don't worry if it sounds a bit weird for now.

Please note that—to tell you the truth—due to cylindrical coordinates' behavior, those coordinates do not define planes but rather cylindrical surfaces around the rotation axis.

Let's now check the tab itself:

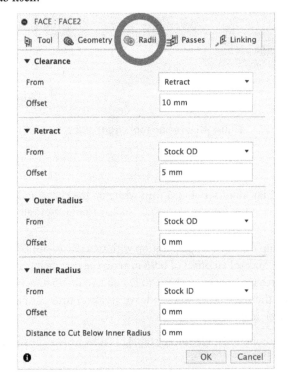

Figure 4.13: Facing's Radii tab

These are the radii we have to work with:

- **Clearance**: The position our tool will move to before and after the operation we are creating. It will be a position where our tool is always safe to retract. Usually, it is the outermost position our tool can assume during our operation.

- **Retract**: As mentioned in *Chapter 1*, between cutting passes, we always need to detach our tooltip from the machined surface (to avoid rubbing) and reposition it for the next pass. With **Retract**, we can specify how far our tool will be detached from the machined surface. The retract range can be set from less than 1 mm all the way up to the clearance coordinate. For extra safety, we may decide to retract our tool after every pass up to the clearance coordinate, but this will only waste time. We usually can just set the tool to retract just a few mm above an already machined surface.

- **Outer Radius**: This is the outermost position our tool will take while cutting. It doesn't have to coincide with our part's outer diameter. For example, what if our stock (which should have an outer diameter of 50 mm) is a bit bigger? We may need to increase the outer radius to be machined to be sure to approach our part correctly.

- **Inner Radius**: This is the innermost position our tool will take while machining. For example, if machining a pipe, there is no reason to let the tool move up until the rotation axis because there is no material to remove there.

We can set the location of all the planes we just reviewed by simply clicking on the **From** drop-down menu:

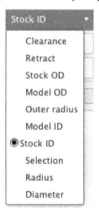

Figure 4.14: Radii selection

As you can see, there are several options to introduce; we are going to mention them all, but they are not difficult to understand:

- **Clearance**: This is a reference to the clearance coordinate (this option won't be available while defining the clearance, of course).

- **Retract**: This is a reference to the retract coordinate (this option won't be available while defining the retract position).

- **Stock OD**: This is a reference to the outer diameter of our stock.

- **Model OD**: This is a reference to the outer diameter of our model.

- **Model ID**: This is a reference to the inner diameter of our model. If our model is not a tube, this must be **0**.

- **Stock ID**: This is a reference to the stock's inner diameter if the stock is hollow or a pipe. If it is not hollow, this value must be **0**.

- **Selection**: This is a surface we can select.

- **Radius**: We can specify a radius by typing a value.

- **Diameter**: We can specify a diameter by typing a value.

Now that we have reviewed all the available options of the **Radii** tab, we can now try to understand what the settings of *Figure 4.13* mean. **Inner Radius** is set to **0** since the stock is not hollow, and we need to machine it up until the rotation axis. Then, **Outer Radius** is set to **0** too, so the outer stock and outer working position are the same (most of the time, you may want to have a slightly bigger working position to compensate for minor errors in the stock). **Retract** is set to be at **5** mm; this means that after a cutting pass, the tool will move 5 mm away from the stock's outer radius. Finally, **Clearance** is set to be **10** mm outward from the retract position.

We can check a 3D view as well to find all these radii set. As you can see, all these positions are highlighted with the corresponding radial coordinate:

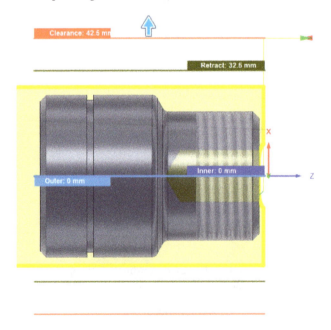

Figure 4.15: Radii position

After reviewing all the key radial positions, we can now move to one of the most important tabs: **Passes**.

Passes tab

Inside the **Passes** tab, not only can we manage all the parameters that regulate the depth of cutting and the number of passes but we can also set the parameters for roughing passes and finishing passes:

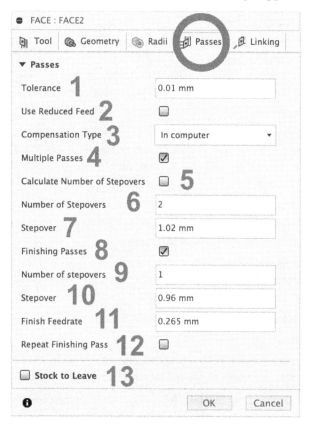

Figure 4.16: Facing's Passes tab

As you may imagine, it is quite dense in content, so let's review it step by step:

1. **Tolerance**: With this value, we can set the **discretization** of complex profiles. Here, we will keep the default value, which is **0.01 mm**; this is fine most of the time, but if you find a segmented shape on your machined part, you may want to reduce this value. Please be careful with this parameter since a direct consequence of a smaller discretization is that a curved toolpath will be described with more points. As a rule of thumb, a 10 times smaller tolerance will lead to 10 times as many points. Each point must be described with a set of coordinates; therefore, the output G-code file may be 10 times bigger. With an unnecessarily small tolerance, we may get a file size of **gigabytes (GB)** instead of **kilobytes (KB)**, and this will slow down the entire process due to limited memory!

> **Note**
> What is discretization? It is a process where we approximate a curve to a set of segments. The smaller the tolerance, the better the approximation and the higher the number of segments created. Something very similar will be covered in *Chapter 17* when dealing with tessellation.

2. **Use Reduced Feed**: With this option, we can specify a smaller feed once the machining operation is approaching its end. Most of the time, it is used when cutting a part from the stock; that's because once the diameter is getting smaller and smaller, the part may start vibrating too much at a high feed rate. In any other scenario, this is something you will rarely have to set. For now, we can leave it unchecked, but we will cover this parameter again in *Chapter 5*.

3. **Compensation Type**: This is the option to shift the tool's origin. We already discussed this when talking about the **Setup** command's **Post Process** tab in *Chapter 2*. As told before, it is a bit beyond the scope of this book, but for now, do not change the default value unless you know what you are doing.

4. **Multiple Passes**: Here, we can specify to Fusion 360 whether we want a single pass or multiple passes. As we already know, for this example part, we plan to perform two roughing passes and a single finishing pass, so we must check this option.

5. **Calculate Number of Stepovers**: With this option checked, Fusion 360 will try to calculate the number of passes required to machine our part. At the moment, we already know the number of passes and every depth of cut, so we will leave this option unchecked.

6. **Number of Stepovers** (for roughing passes): Here, we can specify how many roughing passes we want to generate; in the example, we will set it to **2**.

7. **Stepover** (for roughing passes): With this option, we can set the depth of cut of every roughing pass. For our example, CoroPlus suggested we set it to **1.02 mm** while roughing.

8. **Finishing Passes**: By checking this box, we can specify to Fusion 360 that we want finishing passes to have a better surface quality. This is something we need, as suggested by CoroPlus.

9. **Number of stepovers** (for finishing passes): Here, we can specify how many finishing passes we want to perform. In the example, we will have just **1** pass.

10. **Stepover** (for finishing passes): This value is the cutting depth while performing finishing passes. At the moment, we will set it to **0.96 mm**.

11. **Finish Feedrate**: As you remember, we already set the main feed rate for this facing operation on the **Tool** tab. However, finishing passes always require a different feed, so here we can specify the finishing feed value suggested by CoroPlus: **0.265 mm/rev**.

12. **Repeat Finishing Pass**: As the name suggests, with this option checked, we can repeat another finishing pass that will move along the same toolpath as the first finishing pass. It is mostly used to compensate for part bending; if the part bends, our tool cannot remove the right amount of material because the stock doesn't have the expected shape—therefore, a second identical finishing pass can help remove the bit of material that remained.

13. **Stock to Leave**: If you check this option, you'll be able to specify the amount of material to leave above the machined part. Such an option is most useful when you plan to set two different operations: one for roughing and one for finishing.

Finally, we have completed the **Passes** tab. There is one more tab available called **Linking**, which is used for performance optimization and to regulate how the tool moves between passes. However, this tab is a bit too advanced, and most of the time, you can leave it with default values (unless looking for maximum performance), so for the sake of simplicity, we have to skip this section entirely.

After all of those options, we should now have something like this on our screen:

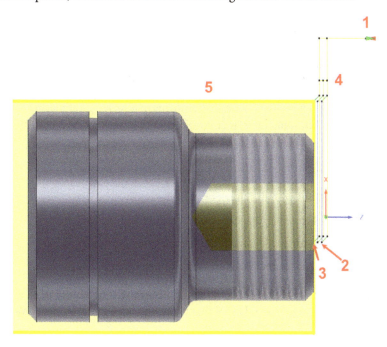

Figure 4.17: Facing complete

As you can see, we have a few lines that describe the toolpath we set with this facing operation:

1. The red arrow is the entry point while the green arrow is the exit point; these points are set on the clearance coordinate and are the first and the last points assumed by our tool.

2. The vertical blue lines are the cutting passes. As you should recall, our facing operation was set with two roughing passes and one final finishing pass. In the screenshot, there are two blue lines detached from the part front, which are the two roughing passes we set. The distance between the lines is the cutting depth, which for our example was 1.02 mm.

3. This vertical blue line is hard to spot, but it is the finishing pass on the front face of our model.

4. These points are on the retract coordinate where our tool moves after every pass (unless otherwise set on the **Linking** tab).

5. This is the remaining stock after this facing operation. As you can see, it no longer protrudes 3 mm from the front of our model.

Congratulations—we just finished our first turning operation!

If you feel a little bored while looking at segments on the screen displaying tool movements, then you are not alone. Fortunately, in the next section, we will find out how to simulate our tool and toolpath for real.

Discovering tool simulation

What is **tool simulation**? The short answer is very simple: it is a procedure where we check whether the result of a turning operation we just set is giving the attended results or not.

In the most ideal scenario, it is just a command that confirms that we are great machinists and that no errors are happening. However, sometimes, we may discover that we still have to tune the operations a little bit more.

While watching the toolpath, we should always check for these four errors:

1. Tool and shank collisions with the chuck or with the stock

2. Entry and exit points located in unsafe areas

3. Tool trying to machine an already machined area

4. Difference between the intended result and the simulated result

It seems that there are so many things that could go in the wrong way that we need to check from our toolpath. Luckily, there is quite a handy tool that takes a great part of the effort. We are about to discover this tool.

Let's start by launching the **Simulate** command:

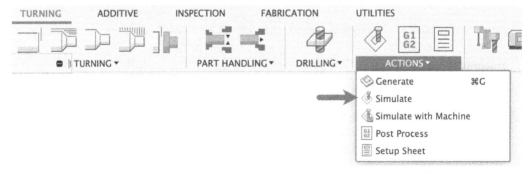

Figure 4.18: Simulate command

Now, we should be presented with the following screen that contains all the information we need to analyze:

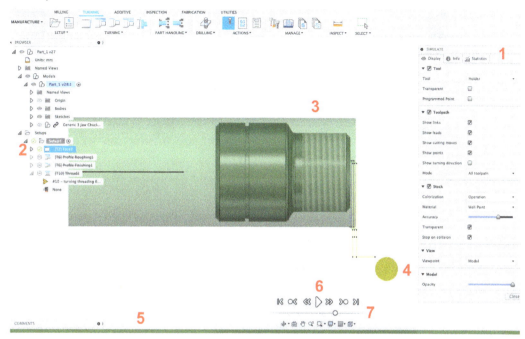

Figure 4.19: Simulate interface

Let's review the most important elements:

1. **SIMULATE panel**: Here, we find lots of options that allow us to tweak the visual appearance of the simulation and get useful data for cost evaluation and tool usage. We will dig deeper into these panels in just a moment.

2. **Current operation**: This is the operation that we are simulating; we may decide to simulate a single operation or the entire setup.

3. **Stock**: This is the stock that we are just about to simulate.

4. **Tool**: This is the tool we used for our first facing operation at its entry location.

5. **Progress bar**: At the bottom of the screen, you may find a bar representing the current progress of the simulation. You can consider it as the progress bar found on any YouTube video or any music player, but showing how much of the cutting is done and how much still has to be completed.

6. **Simulation controls**: Just as with any video player, we can start, stop, and rewind the simulation. If we click on the **Play** button, the tool will move and will start machining our part following the generated toolpath.

7. **Speed cursor**: Since some operations may be very slow, with this cursor, we can increase their speed for visualization purposes.

Just press the **Play** button to check the simulated result:

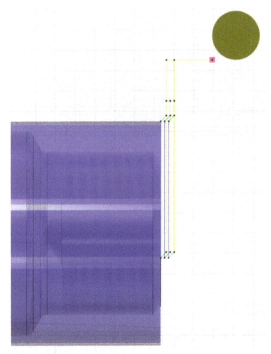

Figure 4.20: Simulation result

As we can find from the screenshot, we can see the transparent stock machined up to the front face of our part and the round insert located on the end of our path. Everything is working as expected!

Let's now review the main **SIMULATE** panel a bit closer. The first tab is called **Display**; here, we can find all the options related to what's visible and what's not during the simulation:

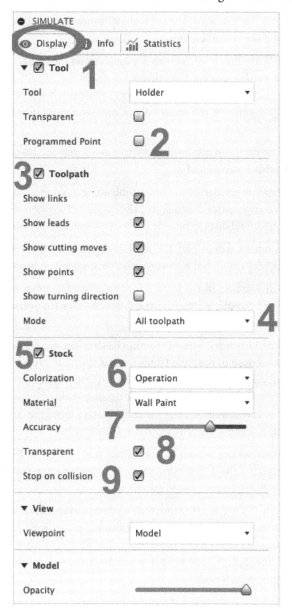

Figure 4.21: SIMULATE panel

As you may expect, there are many settings available, but luckily, most of them are very intuitive; here, we will just have to check the most important ones:

1. **Tool**: With this option, we can show the tool and its holder while simulating the machining operation (you should always enable this option!).

2. **Programmed Point**: With this option, we can display a dot on the simulation tool to show where the origin of the tool itself is located. Note that we can shift this point with the **Compensation** parameter that we mentioned in *Chapter 3* while creating a new tool. During that first encounter, I suggested leaving this point on the tooltip; therefore, we should now find the dot placed on the tip of our tool.

3. **Toolpath**: Here, we can decide to display the toolpath as a group of lines on the screen; I strongly recommend leaving this box checked.

4. **Mode**: With this option, we can decide to display toolpath lines during the entire simulation, only after the simulation is completed, or only before. I always leave this option as **All toolpath**. Basically, changing this parameter will display or hide the tool movements set on the previous option.

5. **Stock**: Enabling this option will display the stock being machined (this is the best part of the simulation since it is very useful but also very satisfying, in my opinion!).

6. **Colorization**: This dialog box allows us to specify how to set colors on the machined faces during simulation; for example, we may have a different color for every different tool used or we can have a different color for each operation.

7. **Accuracy**: Tool simulations work with 3D meshes to simulate machined shapes. Meshes are a type of 3D model composed of multiple triangles that approximates the ideal shape. You can consider the **Accuracy** parameter as the resolution of these meshes (intended as the triangle's density). The smaller those triangles, the higher the detail, but also the heavier the computational task for our computer. If you want smaller meshes, you can move the cursor to the right. Most of the time, however, you can confidently use the default value.

8. **Transparent**: With this option, we can have a transparent stock with the part to be machined inside it. We should always check this option because it allows us to constantly check both the part and the stock.

9. **Stop on collision**: This is the most important option of our simulation, as when checked while simulating, if our tool is colliding with something else, it will immediately stop the simulation and highlight with red the faces that collided. Please note that we may also find minor collisions—some that are hard to spot in other ways; for example, our tool trying to cut a bit beyond its cutting-edge length.

We can now move to the next tab called **Info**; here, we can find all the most important pieces of information on tool position and operation state during the simulation:

Figure 4.22: Info tab

This panel replicates somehow the display or our CNC machine. We will now review the most important points:

1. **Position**: Here, we can find the current position of our tool during the simulation process.

2. **Spindle speed**: This shows the current simulated rotation speed of the chuck.

3. **Feedrate**: This value is the current feed set on the current movement. At the moment, as we can see, it is set to **Rapid**; this is not a numeric value, just the maximum speed our machine can move.

> **Note**
>
> When do we use rapid movement? This is when we need to move our tool from point *A* to point *B* in the fastest way possible—for example, between longitudinal passes (when the tool is moving in the safe retraction area). Rapid movements are never used for cutting purposes since when cutting we have to carefully set the values, considering all the other parameters.

4. **Operation**: Under this subpanel, we can find all the pieces of information related to the type of machining operation and the tool used.

5. **Work offset**: This is a reference to the work offset number we set in *Chapter 2*. It is somewhat difficult to explain, but in short, it is a way to override the default WCS. As you may recall, we simply left the default value of **0**.

6. **Collisions**: This is the total number of collisions encountered up to this point while simulating; it should always be **0**.

7. **Volume**: Here, we can find the mass that will be converted to chip; this is very useful for checking waste material and monitoring costs.

Nice—this tab gave us many useful pieces of information. There is one more tab called **Statistics** to check:

Figure 4.23: Statistics tab

As you can see from the screenshot, here, we can find information related to the total time taken for the machining operations, the total distance covered by our tool, and information related to the number of operations simulated and the number of tool changes with the simulated operations.

Now that we have gone through a full review of the options, we can analyze our simulation results and try to understand whether everything is working fine. As already mentioned, we should answer the following four questions:

1. **Are there collisions?** As found in *Figure 4.22*, the simulation gave us a collision count of 0. Therefore, the most important test is passed!

2. **Are entry and exit points located in a safe area?** From *Figure 4.20*, we can tell that the entry and exit points are located far away from the chuck and from the stock. This test is also passed!

3. **Is the tool trying to cut an already machined area?** To answer this question, we just have to visually check the toolpath or tool movements during a simulation replay. In the following screenshot, I simply stopped the simulation while running the first roughing pass:

Figure 4.24: Simulation in progress

As you can see, all the blue lines are inside the stock (*1*), therefore no cutting movements are wasted cutting air. Also, we can find that the stock is gradually changing shape (*2*), and we can spot the nose radius being impressed on the stock being machined.

4. **Is the resulting shape correct?** This is the simplest question to answer—we can simply check the resulting colors!

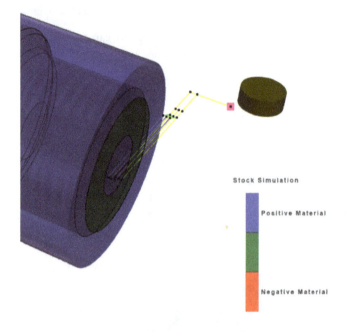

Figure 4.25: Colors

As we can find from the screenshot, the front face of our part is now green, meaning that no extra material is present (it would have been blue) and that we haven't overmachined the geometry, removing too much stock (the color would have been red).

Now, we can be confident that our toolpath is set correctly and that it shouldn't cause any trouble to our lathe or our stock.

Finally, we can focus on converting the machining operations into actual G-code that our machine can understand.

Checking the generated G-code

As you may recall, after setting up all our operations with Fusion 360, we need a postprocessor to generate a file containing instructions that can be read and used by our machine. This set of instructions is called G-code, and every machine may speak a slightly different dialect of G-code.

> **Note**
>
> I would like to highlight the fact that there are hundreds of different postprocessors, each operating with different options and giving different outputs. For this reason, it is impossible to take a deep look at the postprocessing world in the following pages; this would require several books on this subject alone. If you need to understand more about the translation process, I can only suggest you look at the machine documentation or postprocessor developers' documentation.

First of all, we need to launch the **Post Process** command, which you can find here:

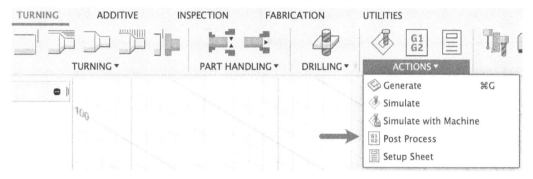

Figure 4.26: Post Process button

After launching the command, we should be presented with a window composed of two subpanels:

- **Settings**
- **Operations**

Let's look at these two panels.

Settings

Settings is the most important and complex panel, where we can specify all the needed information to prepare G-code compatible with our machine:

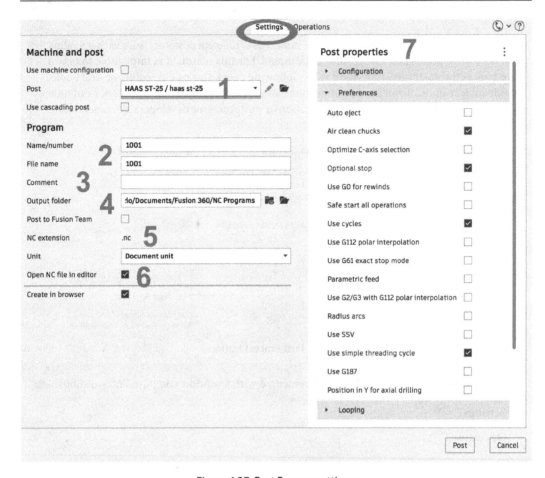

Figure 4.27: Post Process settings

As you can see from the screenshot, there are several interesting options:

1. **Post**: This is the postprocessor that we decided to use. There is a pretty huge list of supported CNC models inside Fusion 360, but in this example, we are going to machine our part on an ST-25 lathe made by HAAS. Therefore, we click on the drop-down menu until we find the right postprocessor.

2. **File name**: This is the filename of the generated G-code file that we will send to our CNC machine.

3. **Comment**: Here, we can specify comments to add to the G-code file. We can add whatever phrase we want—for example, we may use a reference to the part name, its cost, or the author's name.

4. **Output folder**: This is the file path where we are about to save the G-code file.

5. **NC extension**: This is the G-code file extension (note that not every file has the same extension—it may be .gcode or .nc, it may be extensionless, or something else; it depends on your machine).

6. **Open NC file in editor**: With this option, once the G-code file has been exported, it will be opened automatically.

7. **Post properties**: Here, you will find lots of options related to the postprocessor you chose. We must skip this part entirely since it is so largely dependent on the machine and its postprocessor.

This is all we had to know about this first panel, so we can now move to the following one.

Operations

Inside the **Operations** panel, we get a recap of the operations we are about to translate into G-code:

Figure 4.28: Post Process Operations panel

What should we check?

1. **Setup**: This is the setup we are about to translate into G-code; since we may have multiple setups available, we should always check that we are exporting the intended one. As you can see, we can also uncheck machining operations in order to suppress them on G-code output.

2. **Tool**: Here, we can check that the operation is performed using the intended tool.

3. **Work Offset**: This is the work offset reference. Up until now, we found this parameter several times and always ignored it; we will discuss it while talking about milling operations. For now, let's check that the value is **0** so that no WCS override is taking place.

4. **Reorder to Minimize Tool Changes**: With this option, we can try to rearrange operations according to the tool used. This is nice to have, but if you really want to optimize machining time, I suggest you don't use this option but set a proper working order instead.

Once we are ready, we can click on **Post** to create our first G-code file. Let's take a look at the result:

```
%
O01001
G98 G18
G21
G50 S3400
G53 G0 X0.
G53 Z0.


(Face2)
T202
G99
G97
S1333 M3
G54
M8
G18
G0 Z10.
X85.
G50 S3400
G96
S356 M3
G1 Z0.394 F0.491
X65.
X57.828
X55. Z-1.02
X-12.
X-9.172 Z0.394
X65.
Z-0.626
X57.828
X55. Z-2.04
X-12.
X-9.172 Z-0.626
X65.
Z-1.586
X57.828
X55. Z-3.
X-12. F0.265
X-9.172 Z-1.586 F0.491
X85.
Z10.
G97
S1333 M3

M9
G53 G0 X0.
G53 Z0.
M5

M30
%
```

Figure 4.29: G-code

As you can see, the operation that we set up inside Fusion 360 was just converted into G-code that can be used by our machine. Most of the numbers may be hard to understand, but we should try to learn at least the main commands and their effects on the machining process. This may be a challenging task but it will come in handy for future G-code troubleshooting.

If you want to explore this new world of G-code and its functions, the following link is a great starting point: `https://en.wikipedia.org/wiki/G-code`.

Once again, I'd like to highlight the fact that the generated G-code is highly dependent on your CNC and on the postprocessor you use. Therefore, not all commands may be available, and they may not produce the same result!

Congratulations! We finally completed our first facing operation and exported G-code for our CNC—not bad at all! Let's now recap what we have discovered during this chapter.

Summary

At first, we discovered why it is so important to start with a facing operation, then, using Sandvik CoroPlus, we could get all the cutting parameters to implement.

After this introduction, we deeply described all the options for the **Facing** command. This is very important because, as we are about to discover, there are many other turning and milling strategies that we will implement to machine our part, but they will all look and work in the same way.

Once we completed our first operation, we discovered how to simulate our tool and how to check for errors; since we couldn't spot any, we moved forward with postprocessing up to the G-code generation.

It is fair to say that all the efforts we spent on these pages on such a simple facing operation will be worth it time and time again.

Now that we've got an idea of a typical workflow with CAM operations, we can move forward with more complex machining strategies.

In the next chapter, we will keep machining the same example part until its completion. We will discover a bunch of different turning strategies that will allow us to take care of almost any possible shape turnable with a lathe, along with covering longitudinal roughing, contouring finish, threading, drilling, grooving, and parting. As you may imagine, the next chapter will be hugely important!

5

Discovering More Turning Strategies

In this chapter, we will review the most important turning operations available in Fusion 360, in order to implement all the machining needed to complete a somewhat complex design with several different features to be machined. The part we are about to use as an example may look not so difficult, but actually, it hides several challenges to overcome.

Studying this chapter is very important, as we will introduce several new turning operations, which should enable us to machine 99% of the parts we encounter.

In this chapter, we will cover the following topics:

- Turning Profile (Roughing)
- Turning Profile (Finishing)
- Turning Threads
- Drilling
- Turning Groove
- Turning Part

Technical requirements

The main requirement is you need to have read and, for the most part, understood all the previous chapters since we won't explain subjects that we have already dealt with.

Also, as before, we are going to use Sandvik CoroPlus to evaluate the cutting parameters and speed up the process; however, CoroPlus is not strictly needed; for the following pages, you can simply copy the same cutting parameters used in the book.

In this chapter, we will continue with the work we did in *Chapter 4*; therefore, you should have the same setup ready with the facing operation already set.

We will machine the following features out of a stock with a diameter of 55 mm to complete the example part. In the following diagram, you will find all the dimensions you need to define every geometry detail necessary inside the design environment:

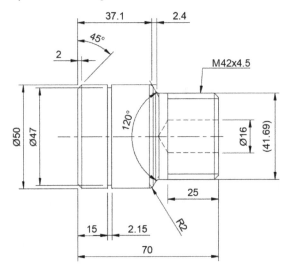

Figure 5.1: Part dimensions

From the diagram, we can spot quite a complex geometry. There is a diameter change at a sloped angle (120°) with R2 fillets, an external thread M42x4.5, and an internal hole with a diameter of 16 mm. There is a groove for a circlip ring, and there are two chamfered edges, 2x45°, one on the front of the part, and one on the rear.

It is now time to start implementing machining operations.

Turning Profile (Roughing)

After the first facing operation we implemented in *Chapter 4*, we should start by setting up a roughing operation to remove most of the material from the stock. One of the best commands for roughing our part is called **Turning Profile Roughing** and can be found among all the other turning commands:

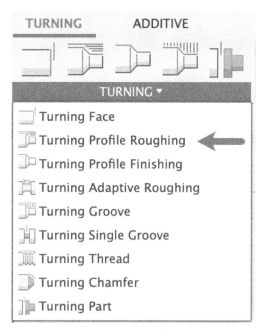

Figure 5.2: Turning Profile Roughing

This turning operation will generate longitudinal passes, and it is useful to use when removing high amounts of material with powerful tools operating at a high cutting power. Before clicking on the command icon, we will choose the cutting tool we plan to use.

For this roughing operation, I decided to use the following ISO-coded insert: **CNMG160616**.

> **Note**
>
> Why did I choose to use this tool (**CNMG160616**)? Because it is good for both roughing and finishing, and most importantly, it doesn't have accessibility issues like the round insert we used for facing.

The insert will be installed on the following shank I created:

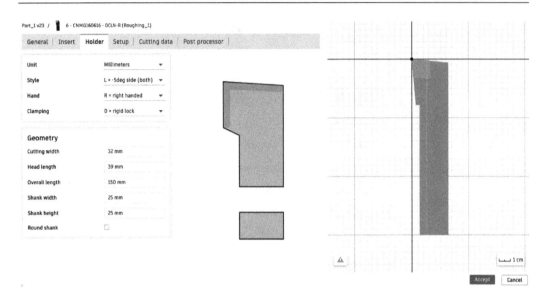

Figure 5.3: Tool Shank

For now, you can simply copy and paste the values in order to follow the example operation. Now that we have a proper tool at our disposal in the tool library, we can start the **Turning Profile Roughing** command. We should now be presented with the typical panel with multiple tabs where we can specify all the cutting settings.

The Tool tab

Inside the **Tool** tab are the options that allow us to choose the cutting tool and its cutting parameters:

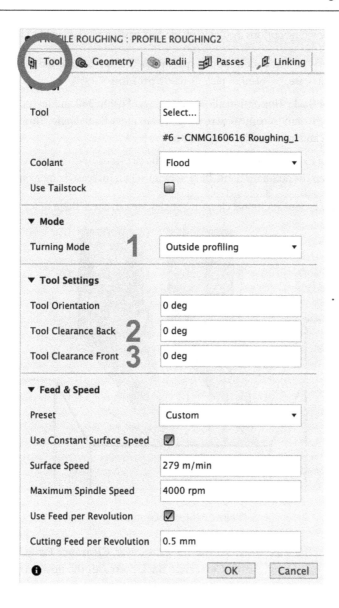

Figure 5.4: The Roughing Tool tab

This tab is very easy to understand, and you may have noticed that **Surface Speed** should be set to **279 m/min**, **Maximum Spindle Speed** to **4000 rpm**, and **Cutting Feed per Revolution** to **0.5 mm**. All these values were calculated using CoroPlus.

Let's now focus on a few new options available inside this first panel (labeled in Figure 5.4):

1. **Turning Mode**: With this option, we can specify whether it is an internal or an external turning operation; in our case, we need to pick **Outside profiling**.

2. **Tool Clearance Back**: This option allows us to cheat Fusion 360 and let it think the tool has a bigger nose angle than in reality, increasing the auxiliary flank angle. This is used to produce a safety margin and avoid rubbing against the part.

3. **Tool Clearance Front**: This is very similar to the previous option, but instead of specifying a bigger angle on the auxiliary flank, here, we can set an increased angle on the main flank.

If you feel a bit confused about additional clearance angles, we can check them in the following diagram:

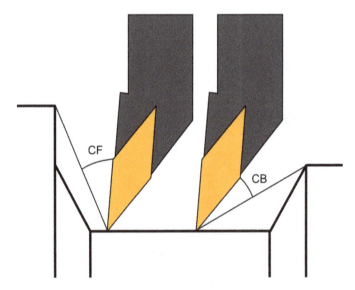

Figure 5.5: Clearance angles

As you can see, both options increase the nose angle; however, **Clearance Front** increases the nose angle by rotating the main flank, while **Clearance Back** increases the nose angle by rotating the auxiliary flank of the insert. Since the faces we are machining are not very deep and don't have a complex profile, we can leave the default value of **0** degrees on both. A value of **0** won't have any effect on toolpath generation.

The Geometry tab

Inside this tab, we can specify the boundaries for our machining operation.

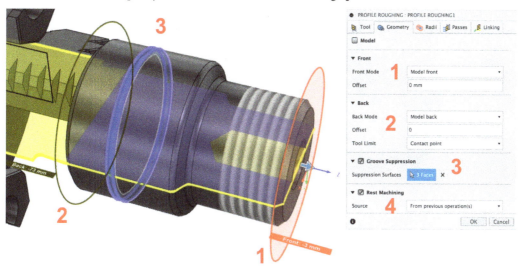

Figure 5.6: The Roughing Geometry tab

Only faces inside the specified boundaries will be considered for toolpath generation, while everything else won't be machined. The main goal of this panel is to set the position of the two boundary planes called **Front** (*1*) and **Back** (*2*). As you can see from the screenshot, we can set the position manually by dragging the two planes directly from the 3D view or by specifying the parameters inside the drop-down menus.

> **Note**
>
> It may be a good idea to increase the machined length on the back a little bit if you plan to cut the part from the chuck.

Just a couple of things here to notice. We will machine the groove later with a dedicated operation, so we can remove all those faces with **Groove Suppression** (*3*). With this option, Fusion 360 won't calculate them when generating the toolpath.

Also, since this is not the first operation of the setup, we must tick **Rest Machining** (*4*). With this option ticked, the tool will be aware of areas that have already been machined by previous operations, and it will skip additional machining. In the example, we already machined the front side of our model, so there is no need to machine it again!

The Passes tab

As the name might suggest, inside this tab, we can specify details that will tweak the cutting passes of the turning operation. This tab is one of the most important since, as we are about to discover, it can drastically change the cutting strategy:

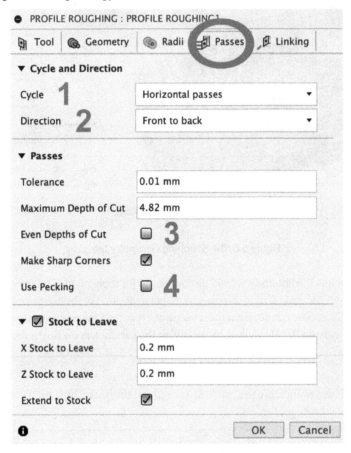

Figure 5.7: The Roughing Passes tab

Inside this tab, we can specify the depth of cut and set additional stock to leave around our shape, but there are also a couple of interesting settings we should review in detail (labeled in Figure 5.7):

1. **Cycle**: With this option, we can specify whether the tool should cut in horizontal passes or in radial passes (such as when facing). Most of the time, we want **Horizontal passes**.

2. **Direction**: This is used to specify the direction of our tool when cutting. For example, it may cut while getting closer to the chuck (**Front to back**) or while moving away from it (**Back to front**). We can also specify to cut **Both ways**. Note that you have to choose the direction according to your tool and shank shape; setting the wrong cutting direction may cause stock collisions.

3. **Even Depths of Cut**: With this option, we force Fusion 360 to use, as much as possible, a constant cutting depth. Even depths of cut are typically used for a constant cutting edge engagement when performing several roughing passes. In our example, we just need one pass where the diameter is bigger and two passes where the diameter is smaller; therefore, we can leave the option unticked.

4. **Use Pecking**: With this box ticked, our tool will cut a bit in the forward direction, then it will move backward and then forward again. This option is mostly used as a chip-breaking strategy, which we don't have to implement here.

As you may have noticed, I decided to leave **0.2 mm** of stock around my part since I plan to add an additional finishing pass to the part. The value of **Maximum Depth of Cut** is set at **4.82 mm**, and it was calculated using CoroPlus. For our example, we can leave everything else on the default settings.

This is all we need to know about the **Passes** tab, and the next tab is called **Linking**. Unfortunately, as mentioned in the previous chapter, it is still out of our grasp. However, don't worry, as most of the time, you can leave it with the default settings; therefore, we can consider our first roughing operation completed!

Simulation

We already discovered how to check the tool simulation in *Chapter 4*; therefore, we will skip all the options and jump straight to the results if we want to check the machined geometry with a simulation. We should find the following result:

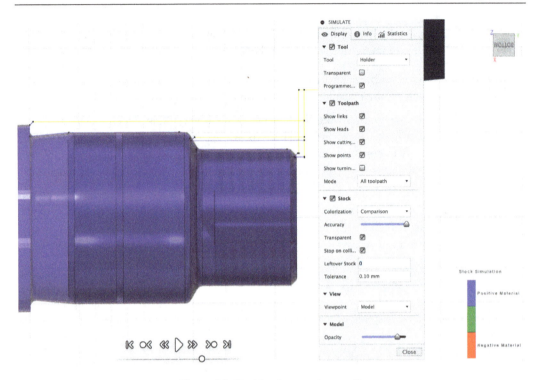

Figure 5.8: Machined geometry result

There are a couple of things to highlight from the screenshot:

- As expected, the groove was not picked for machining since we suppressed the feature
- We machined a bit beyond the back of the part, as discussed on the **Geometry** panel
- The stock is blue, meaning that there is extra material to remove for the next finishing operation (we left a layer of 0.2 mm all around)

For now, everything is working just as expected. In the following section, we will add the finishing pass and will remove the additional stock left around our geometry.

Turning Profile (Finishing)

After we have implemented the first roughing operation, it is a good idea to set a final finishing pass to remove only a tiny bit of material and achieve a good surface quality.

Fusion 360 comes with a handy operation called **Turning Profile Finishing**, which we can find among all the other turning operations in *Figure 5.2*:

This is a contouring operation that will generate a toolpath parallel to the part profile. Before launching the command, we should know which type of tool we plan to use and which cutting settings will allow us to have the target surface quality we want.

For this operation, we can still use the same tool and shank used in the previous operation. The insert (**CNMG160616**) can be used both for roughing and for finishing, in fact. Once again, we will let CoroPlus calculate the optimal cutting parameters for the insert and simply copy and paste them. Let's now launch the command!

The Tool tab

As we should already know, inside this tab, we can specify the tool to be used and its main cutting parameters:

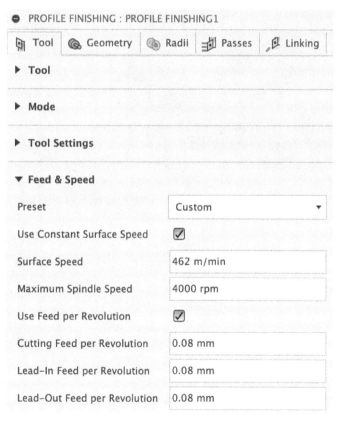

Figure 5.9: The Finishing Tool tab

Here, CoroPlus suggested a much lower **Cutting Feed per Revolution** in contrast to the roughing operation; here we have **0.08 mm**, while previously, we had **0.5 mm**, almost 10 times smaller!

Cutting speed (sometimes referred to as **Surface Speed**) is set at **462 m/min** while the **Maximum Spindle Speed** is set to **4000 rpm** since it is the maximum value my lathe can spin.

The Geometry tab

Inside the **Geometry** tab, we can specify the working area as well as decide to exclude selected faces from toolpath generation. This is the same concept found in the previous roughing operation; since it is very similar to what we already explained, we will review this panel more quickly:

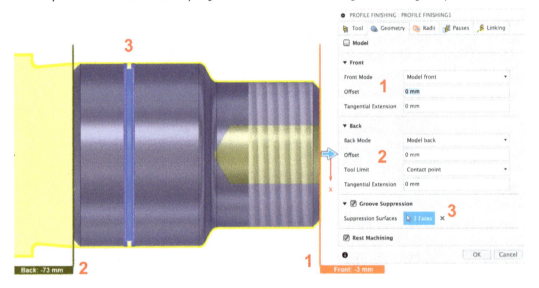

Figure 5.10: The Finishing Geometry tab

Here, we have to set the boundaries for the finishing operation by setting the **Front** plane (*1*) and the **Back** plane (*2*). Since we need to remove the extra stock left during the previous roughing operation, we need to machine the entire part length; therefore, we have to set the cutting boundaries to **Model front** and **Model back**.

Once again, we still have to remove the groove from the machined geometry with the **Groove Suppression** feature (*3*). We will machine the circlip groove with a dedicated operation later in this chapter.

The Passes tab

Inside this tab, we can find all the parameters to affect cutting passes:

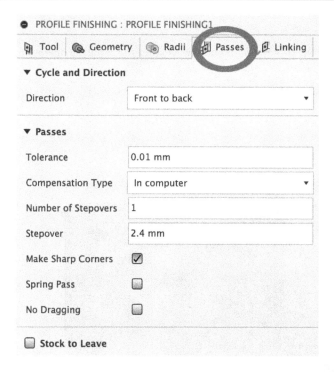

Figure 5.11: The Finishing Passes tab

These are the most notable options we have to mention:

- **Number of Stepovers**: Here, we can specify how many cutting passes we want to perform. In our example, we simply need to remove a material layer of 0.2 mm; therefore, we can remove it with a single pass.

- **Stepover**: With this value, we can set the maximum cutting depth for our tool. At the moment, there is a default value of **2.4 mm**. In reality, the tool will never cut at that depth since only 0.2 mm has to be removed.

- **Spring Pass**: With this ticked, we can decide to perform a second pass identical to the first one. This is typical of finishing passes; it is used to remove the material that the first pass couldn't remove due to part bending. However, most of the time, you won't need it if you set the cutting parameters correctly.

- **No Dragging**: The goal of this option is again to improve the final part tolerances and finish by applying less negative pressure on the tool. Ticking this option will prevent the tool from cutting while pulling. Note that this option may increase machining time significantly.

Simulation

It is now time to check the simulation to find out whether the resulting operation is working:

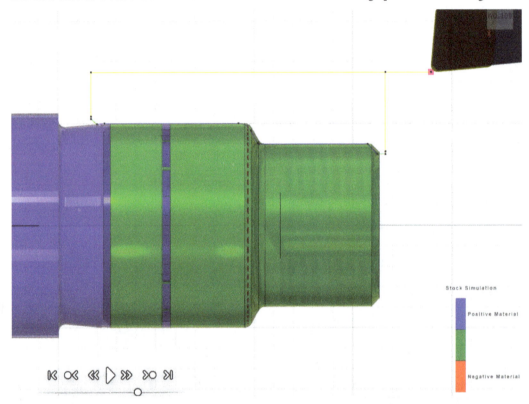

Figure 5.12: Simulation results

The picture looks almost identical to the previous results with the roughing passes. However, there is only one notable exception: the stock is now green, meaning that there is no extra material left on the part (it would be blue if there were) and that it was not over-machined (it would be red if it had been).

Now that we have introduced the main finishing strategy for turning, we can start with something a bit spicier; let's try threading, but be careful — threading is one of the most difficult operations.

Turning Threads

In this section, we are about to discover how to machine threads with our lathe, but before jumping directly into the turning command that allows us to machine threads, we need to recap thread theory and check our thread specifications.

As you may recall, our example part features a thread starting from the front face. The first thing we will do is check the thread specifications. The thread we have to machine is an M42x4.5, a standard ISO thread with a coarse pitch.

All ISO screw threads are ruled by the following diagram:

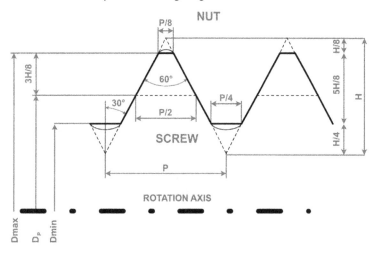

Figure 5.13: ISO thread

Now that we can check the thread drawing, we can explain the parameters we can spot inside the diagram:

- **P**: This is the thread pitch. For a coarse M42, the pitch is 4.5 mm.

- **Dmax**: This is the screw's maximum outer diameter. For an M42, this value must be between 41.937 mm and 41.437 mm.

- **H**: This is the thread height measured between sharp corners:

$$H = \frac{\sqrt{3}}{2} * P$$

In our example, H is 3.897 mm.

- **Dp**: This is the pitch diameter, and it is equal to the following formula:

$$D_P = D_{max} - \frac{3\sqrt{3}}{8} * P$$

In our example, D_P is 39.014 mm.

- **Dmin**: This is the internal diameter for our thread. It can be calculated with the following formula:

$$D_{min} = \mathrm{D}_{max} - \frac{5\sqrt{3}}{8} * P$$

For our thread, this value is 37.066 mm.

Let's keep these values in mind since we will need them later for the threading setup. The next step is to determine the cutting parameters for the operation we are about to create; once again we will use CoroPlus and use the suggested parameters:

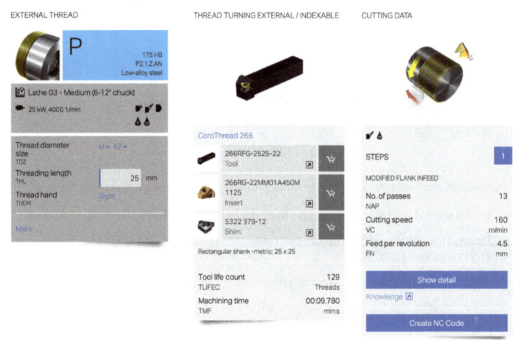

Figure 5.14: Cutting parameters

As we can see, the suggested insert is **266RG-22MM01A450M**. We will skip the part about how to create this tool since we should already be experts on this subject; instead, let's analyze the suggested cutting parameters, and we can find something interesting.

First of all, as expected, we can see a **Feed per revolution** value of **4.5 mm**/rev, a suggested **Cutting speed** of **160 m/min**, and **13** machining passes. Up until now, we've never had to set so many cutting passes!

Such a slow cutting speed and so many passes should suggest to us that threading can be quite a power-hungry task. In the following pages, we are about to discover why threading is such a complex operation.

Now that we have all the numbers we need to machine and measure our thread, we can finally launch the **Turning Thread** command that can be found in the drop-down menu shown in *Figure 5.2*. As always, you will find yourself in the **Tool** tab.

The Tool tab

Inside the **Tool** tab, alongside the tool we plan to use, we also have to specify the type of threading we want to machine and the spindle rotation speed. We have already covered the tool we are about to use; therefore, let's focus on the other settings:

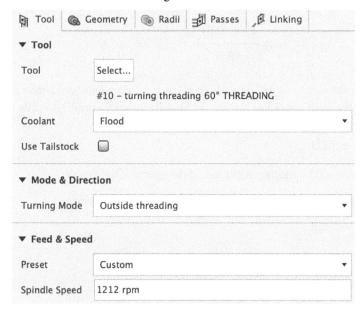

Figure 5.15: The Threading Tool tab

As you can see from the screenshot, this tab is quite similar to the **Tool** tab for the other operations we have already set up; there are just a couple of differences and hidden troubles I'd like to highlight.

With **Turning Mode**, we can set whether we are performing **Outside threading** or **Inside threading**. To keep things simple, let's just say that **Outside threading** will machine something like a screw, while **Inside threading** will result in something like a nut. Our part looks like a screw, so we can leave the default value, **Outside threading**.

The only other value we need to set is the **Spindle Speed**, which is expressed as **rotations (or revolutions) per minute**. CoroPlus gave us a cutting speed expressed in **meters per minute (m/min)**, so we need to convert it to rpm. Let's see how to do this. As you may recall from the first chapter, cutting speed can be calculated using the following formula:

$$V_c = \frac{\pi n D}{1000}$$

Where n is the rotation speed expressed in rpm, D is the diameter expressed in mm, and V_c is the cutting speed expressed in m/min. Using that formula, we can find the value of n, which is equal to the following:

$$n = \frac{V_c * 1000}{\pi D} = 1212\ rpm$$

Once again, the equations have proved handy in solving a simple unit conversion issue! Now that we have calculated the exact rpm value, we can set **Spindle Speed** accordingly inside the **Tool** panel.

The Geometry tab

Inside this tab, we can set the starting position and the length of the thread:

Figure 5.16: The Threading Geometry tab

With **Thread Faces**, we can select the cylindrical surfaces where we want to machine our thread. There is a useful option called **Confinement** that we should enable, allowing us to specify the machining boundaries for this operation.

With **Frontside Stock Offset**, we can specify an additional length on the front of our part to be threaded; this is always a good idea, as the tool will start threading a bit before the part begins, and this ensures that the entire length is machined properly.

With **Backside Stock Offset**, we can specify an additional length after the thread face has ended.

If we enable the **Apply Back From Front** option, we can change the way **Backside Stock Offset** is calculated: instead of being calculated from the back of the part, it is measured from the front plane. It may sound more complex than it is, so take a look at *Figure 5.17* for extra clarity:

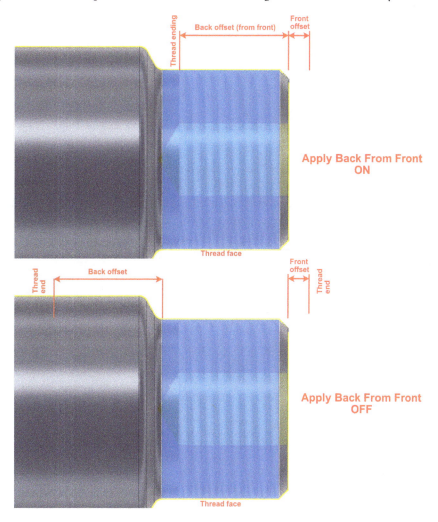

Figure 5.17: Apply Back From Front

As you can see from the picture, with **Apply Back From Front** enabled, we can only machine a portion of the thread face, while with the option disabled, the back offset is set from the end of the thread face.

In our example, the thread is 25 mm long, and I want a safety margin of 5 mm on the front of our thread, so I set the values accordingly.

The Radii tab

Inside the **Radii** tab, we can specify the distance our tool has to retract between cutting passes; most of the time, just a couple of millimeters is sufficient:

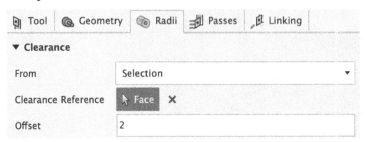

Figure 5.18: The Threading Radii tab

In the example, I set the **From** option to **Selection**, selecting the threaded face of the part. This way, the clearance will be calculated from the outer diameter of the thread. For **Offset**, we can add **2** mm as a safety margin.

The Passes tab

This tab is fundamental because it is where we have to input the values we calculated using the formula; even a small error here will lead to a bad result:

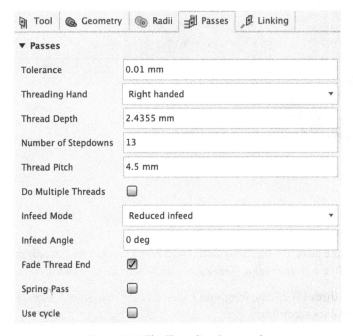

Figure 5.19: The Threading Passes tab

There are several options we must specify with extreme precision:

- **Threading Hand**: This option affects the way we tighten or loosen our thread; The **Right handed** threads get tightened with a clockwise rotation, while the **Left handed** threads get tightened with a counterclockwise rotation

- **Thread Depth**: This value is about how far the tool will enter into the stock during the final machining pass and it can be calculated with the following formula:

$$\frac{D_{max} - D_{min}}{2}$$

So, for our example, the formula is as follows:

$$\frac{41.937 - 37.066}{2} = 2.4355mm$$

- **Number of Stepdowns**: With this value, we can specify how many passes to perform during the threading operation. CoroPlus suggested we perform **13** passes, so we simply input the same value in the box.

- **Thread Pitch**: As previously mentioned, this value derives from the thread specifications; in our example, we are machining an M42x4.5; therefore the pitch is **4.5 mm**.

- **Do Multiple Threads**: This option allows us to perform multiple threads. There are certain types of screws with multiple start threads, which are typically used for linear motion control. In this example, we are performing a standard bolt thread, so we shouldn't tick this box.

> **Note**
> You can discover more about multi-start threads in this article: `https://blog.igus.co.uk/difference-between-single-and-multi-start-threads/`.

- **Infeed Mode:** This setting manages the working load on our cutting edge, adjusting the depth of cut accordingly. There are three different options, as listed here:

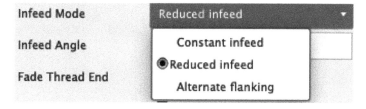

Figure 5.20: Infeed Mode options

- **Constant infeed**: This is the most traditional mode where every cutting pass is performed at the same cutting depth. However, this is the worst mode because, with constant cutting depth, the working load on the cutting edge increases quite a lot from one pass to the next.

- **Reduced infeed**: With this option, we reduce the cutting depth at every pass to have a constant cutting-edge load.

- **Alternate flanking**: This is the most advanced strategy. Not only does it decrease the cutting depth at every pass, but it will also machine one side of the thread and then the other.

You can be confident in both **Reduced infeed** and **Alternate flanking**, but you should always avoid **Constant infeed**.

- **Fade Thread End**: With this option, we can stop the threaded area smoothly without a sudden interruption. You should always pick this option if performing a limited-length thread, such as our example here.

- **Spring Pass**: This is similar to what we found before when reviewing the **Turning Profile (Finishing)** strategy; it is a second pass that aims to compensate for any excess material due to part bending. In this case, the stock is quite thick and short; therefore, we don't have to worry about bending.

- **Use cycle**: This is an advanced option that can work in tandem with the Post processor to reduce the amount of G-code generated (unfortunately, it is way beyond the scope of this book, so we won't explore this topic).

Simulation

Now that we have set all the important parameters for our thread, we can run a simulation to check the result:

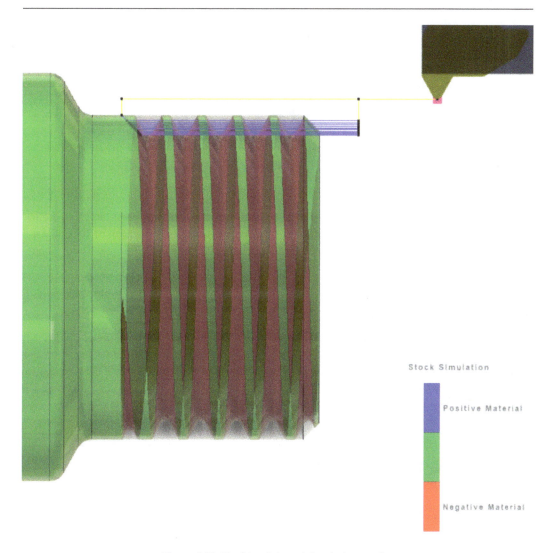

Figure 5.21: Machined thread simulation results

The thread looks fine, but you may have noticed that the machined faces are red, which typically means that too much material was removed. Normally this would be an error but not in this case. Every time we create a threaded part, we never model the thread itself (it would be too difficult and time-consuming); we simply apply a cosmetic thread on the cylinder using a texture. So, with this warning, Fusion 360 is simply telling us that we went inside the modeled cylinder; therefore, we can safely ignore this warning.

I think it is now time to move to the next operation, turning drilling!

Drilling

We haven't mentioned this yet, but a lathe is also capable of drilling. When drilling with a lathe, the drill bit doesn't spin; it simply moves against the stock, and it is the part that is spinning.

There is one limitation, though: since we use the chuck rotation to remove material, the hole axis must be coincident with the rotation axis. There is no workaround to this limitation; it is the way drilling works with a typical lathe. If our part features multiple holes, we may need to use a milling operation after processing the stock on our lathe.

Another option would be a multi-spindle machine that can perform both milling and turning operations.

In the following pages, we are about to find out how to set up a simple drilling operation. Please note that there is not a dedicated drilling command for turning; the same command can be used when setting up both turning and milling operations. This is the reason why we will encounter few options inside the command that are not applicable to a lathe.

Let's get back to our example part. It features a 16 mm diameter hole on the front face, and in the following pages, we will try to create this hole.

> **Note**
>
> If we had to create an internal thread on our part, we could reuse the **Turning Threads** command and create an internal operation. However, if our thread hole is too small, we may even decide to perform a drilling and taping operation with a standard bit.

First of all, locate the **Drilling** command, as shown in the following screenshot:

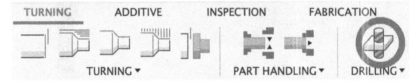

Figure 5.22: The Drilling command location

As usual, the command is divided into multiple tabs that we are about to analyze one by one.

The Tool tab

The **Tool** tab is quite similar to what we have already encountered several times:

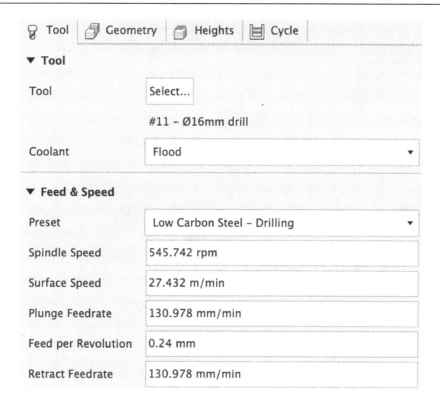

Figure 5.23: The Drilling Tool tab

As you can see, we can specify the drill bit to use and the cutting parameters. Luckily, Fusion 360 comes with lots of drill bits; therefore, most of the time, we should be able to find the diameter we need inside the built-in tool library. In our example, we will pick a drill bit that is 16 mm wide.

This time we don't really need to use CoroPlus since there is a very handy **Preset** drop-down menu that allows us to specify the cutting parameters according to the stock material; in this case, we will use the default preset, **Low Carbon Steel - Drilling**.

The Geometry tab

Inside this tab, we can specify where we want to drill the holes. Please note that in turning, as already mentioned, we can drill along the rotation axis only; therefore, all the options managing multiple holes can be ignored; we will introduce them, anyhow, just for a better understanding of the command:

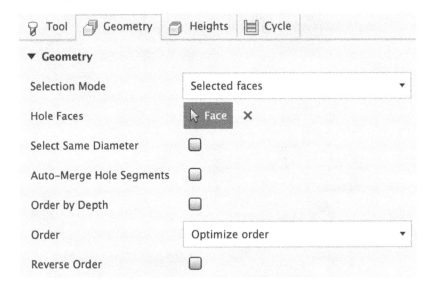

Figure 5.24: The Drilling Geometry tab

Selection Mode allows us to specify different ways to pick holes to be machined inside the part. When setting up a turning operation, we can drill a single hole along the rotation axis; therefore, we shall use the option called **Selected faces**, which allows us to pick the faces to be drilled.

However, with a milled part, we may have hundreds of holes to drill; therefore, instead of selecting them all one by one, we can tick the option called **Select Same Diameter**. With this option ticked, our operation will automatically drill every hole with the same diameter as the first one selected.

Then the next options, **Order by Depth**, **Order**, and **Reverse Order**, allow us to sort our drilling operations.

The following tab is called **Heights**, but we won't review it since it is very similar to the **Radii** tab that we have already seen multiple times. It is used to set the main working positions for our operation, such as the clearance height and so on.

> **Note**
> Why is it called **Heights** instead of **Radii**? The answer is quite simple: drilling is not considered a turning operation inside Fusion 360; therefore, it shows an interface closer to a milling operation.

The Cycle tab

This tab is quite important, as we can differentiate between drilling operations:

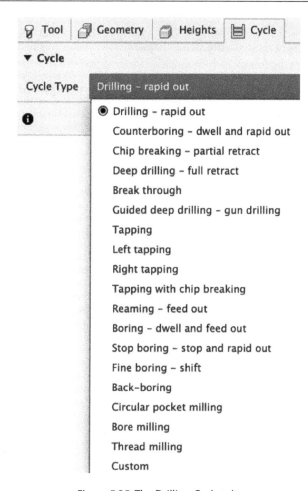

Figure 5.25: The Drilling Cycle tab

Though it's small, you can see there are several options to choose from to select the type of operation we intend to set up. For our example part, we just need a simple **Drilling – rapid out** operation, which is the default setting for standard blind holes. But from this panel, we can select other settings, such as the **Tapping** and **Counterboring – dwell and rapid out** operations.

> Note
>
> Tapping is an operation that machines a thread inside a hole with a special tool called tap for threading. Counterboring is an advanced drilling operation that machines the housing for the head of a screw.

Unfortunately, we won't dive deeper into these operations because this chapter is centered on turning and because doing this would require a dedicated chapter; however, feel free to do your own research if you're interested in learning more.

After this fast introduction to drilling with a lathe, we can move to the next cutting strategy we need to complete the example part: **Turning Groove**.

Turning Groove

In our part, there is a circlip groove that we still have to machine. As we can spot in *Figure 5.2*, there are two different types of grooving strategies we can use: **Turning Single Groove** and **Turning Groove**.

What is the difference between the two? With **Turning Single Groove**, we can specify an edge where we want to perform a groove. And that's it; almost no other option is available. On the other hand, **Turning Groove** is a much more powerful command; not only can it be used for groove cutting, but it can also be used as a roughing strategy instead of typical longitudinal machining.

Before starting the command, we already know the tool we have to use and its cutting parameters. The tool we are about to use is coded as: **N123T3-0150-0000-GS-1125**, which is a specific tool for grooving. For additional information on this tool, you can find the datasheet at the following link: `https://www.sandvik.coromant.com/en-gb/products/pages/productdetails.aspx?c=n123t3-0150-0000-gs%201125`.

As always, we can leave it to CoroPlus to produce all the equations. For the example, CoroPlus is suggesting a cutting speed of 242 m/min, a cutting feed of 0.05 mm/rev, and a chuck rotation of 1,640 rpm.

Unfortunately, Fusion 360 has limited support for grooving tools; therefore sometimes, when creating a new tool from scratch, we can only set the most important dimensions, meaning that it will not look like the real tool, but it will still work just as well. In the following screenshot, we can find a simple tool that will pretend to be the real one:

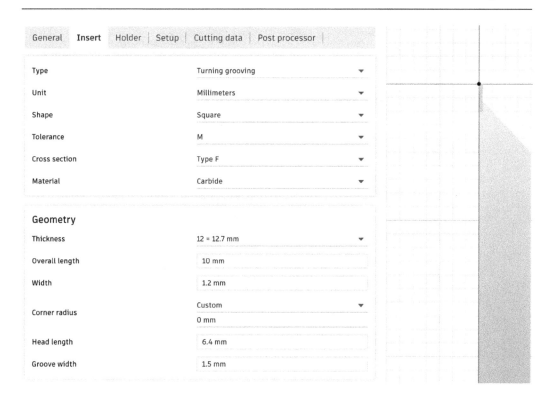

Figure 5.26: The grooving tool

As you can see, it is quite a crude tool model, but it will work perfectly well for CAM purposes. For our scenario, we only have to look at the following three parameters:

- **Groove width**: This is the minimum width this tool is able to cut. In our example, the groove has a width of 2.15 mm. Therefore, we can use any tool with a smaller **Groove width**.

- **Corner radius**: This is the radius machined at the bottom of the groove. In our example, there is a sharp edge; therefore, a radius of **0 mm** is mandatory.

- **Head length**: This is the maximum groove depth the tool can machine.

Now that we have the right tool in our library, we can finally launch the **Turning Groove** command and start setting up the operation.

The Tool tab

As usual, the first tab we need to look at is a rather typical **Tool** tab that we should be already familiar with:

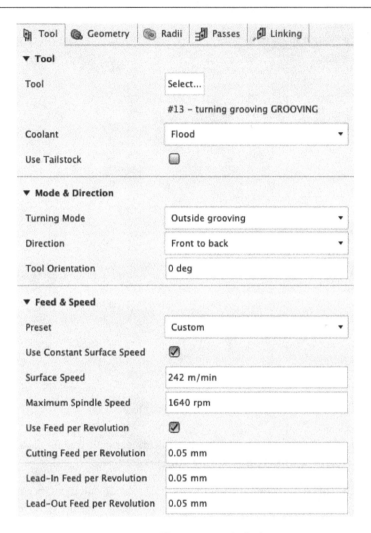

Figure 5.27: The Grooving Tool tab

With **Turning Mode**, we can specify whether we want to perform **Outside grooving** or **Inside grooving**.

The next option is called **Direction**, and it is used to specify the cutting direction of the tool; we have three options to choose from:

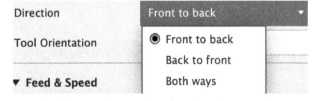

Figure 5.28: Cutting Direction drop-down menu

This is the same key idea as we found in the second point of *Figure 5.7*:

- **Front to back**: Here, the tool will start cutting from the front of our stock, one cutting pass after the other, until it reaches the rear of our part

- **Back to front**: Here, the tool starts from the rear of the part and moves toward the front of the stock

- **Both ways**: With this option, the tool can cut back *and* forth

In order to choose correctly, we always have to consider our tool geometry and how the cutting edge is oriented. For our example tool, it doesn't really matter because it has two cutting edges, but we have chosen **Front to back** (even if it could cut both ways, I would still choose this option).

The Geometry tab

As the name suggests, inside this panel, we have to specify geometry-related parameters and the working area:

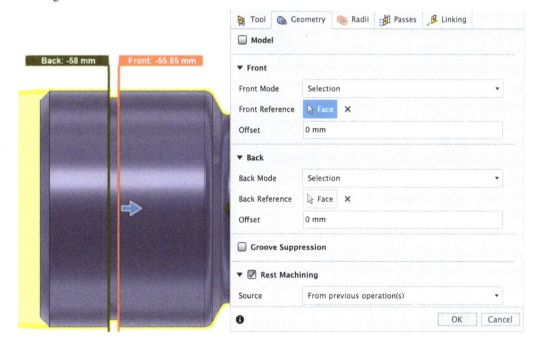

Figure 5.29: The Grooving Geometry tab

We can limit the working area of our groove by simply using **Front Reference** and **Back Reference**. By selecting the two flanks of our groove this way, the tool won't cut anywhere else but inside the groove itself.

Once again, we find the **Groove Suppression** option already encountered in *Figure 5.6* and *Figure 5.10*. Of course, at the moment, we don't need to suppress this feature since we are trying to machine it, so we can leave **Groove Suppression** unchecked.

Another option already seen but worth mentioning is **Rest Machining**. With this option selected, Fusion 360 won't try to remove stock material where it has already been removed with previous operations. This is something we shall apply since, with previous operations, we already removed 2.5 mm of stock above the groove during the roughing passes.

The Passes tab

As usual, the **Passes** tab has several options that we need to properly understand to master the command:

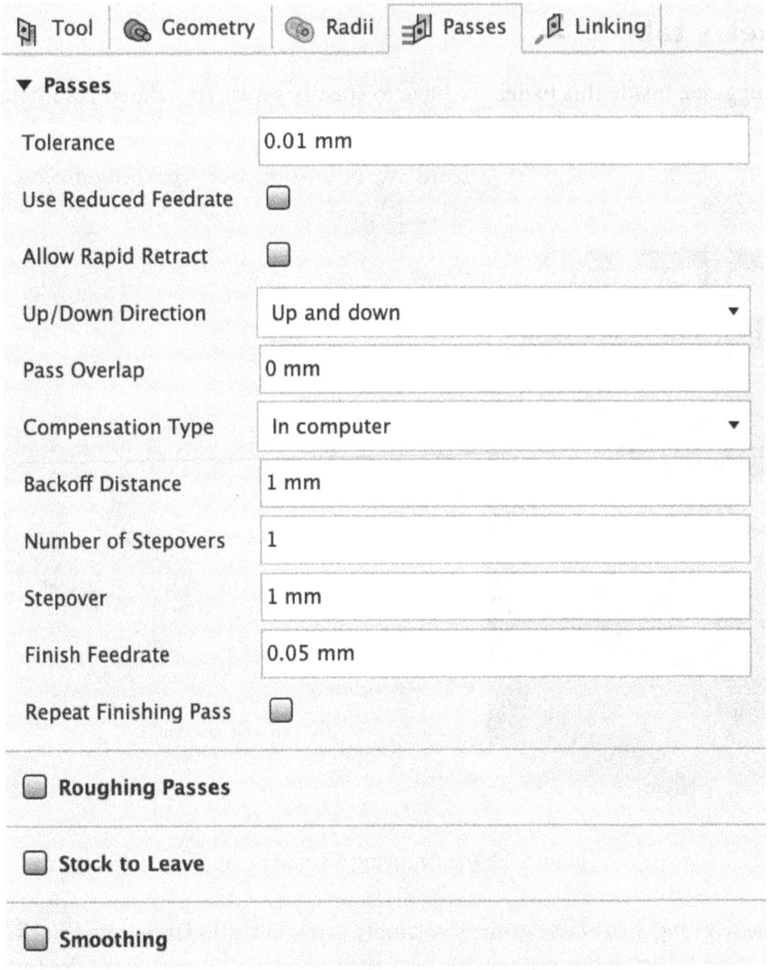

| Tool | Geometry | Radii | Passes | Linking |

▼ **Passes**

Tolerance	0.01 mm
Use Reduced Feedrate	☐
Allow Rapid Retract	☐
Up/Down Direction	Up and down ▾
Pass Overlap	0 mm
Compensation Type	In computer ▾
Backoff Distance	1 mm
Number of Stepovers	1
Stepover	1 mm
Finish Feedrate	0.05 mm
Repeat Finishing Pass	☐

☐ **Roughing Passes**

☐ **Stock to Leave**

☐ **Smoothing**

Figure 5.30: The Grooving Passes tab

As we can see, there are some elements already found in other turning operations, for example:

- **Roughing Passes**: This allows us to set multiple passes before a final pass. In our example, we will use a single pass; therefore, we shall leave this unchecked.

- **Stock to Leave**: We can use this option if we plan to implement an additional finishing pass over the machined geometry. For our example, we won't enable this since we will machine the groove in only one pass.

- **Use Reduced Feedrate**: This can be useful if our groove is very deep. Let's dig more into this option:

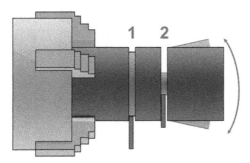

Figure 5.31: Groove depth comparison

As you can see in the diagram, the first groove is not very deep; therefore, it doesn't compromise the stock strength; the part is not likely to bend there even with a high radial cutting load. On the other hand, the second groove is very deep, so the part may start vibrating and get damaged. A smaller feed will help generate a lower radial load when cutting close to the rotation axis. This same concept will be found when cutting our part from the chuck in the *Turning Part* section.

- **Up/Down Direction**: This is an option that allows us to specify the cutting direction. We can force our tool to cut both when plunging inside the part and when retracting (**Up and down**), to cut only when plunging (**Down only**), or only when retracting (**Up only**). Changing the cutting direction may help us to reduce the radial load on the stock.

- **Pass Overlap**: This is an option that forces our tool to partially overlap the previously machined area to be sure that all the material has been removed. It is similar to the **Spring Pass** option we have already encountered.

- **Stepover**: With this, we can set the maximum plunge depth for a single grooving pass. While with **Number of Stepovers**, we can set how many passes we want to perform. Again, this is very similar to the longitudinal roughing passes we have already covered.

After looking at the **Passes** tab, as usual, we will leave the **Linking** tab behind. Now our first grooving operation should be complete, it is now time to review the result with a simulation.

Simulation

Finally, our part is almost completed. I know that watching all the cutting operations being simulated can be somehow hypnotizing; however, we should only focus on the resulting geometry:

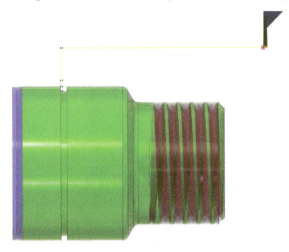

Figure 5.32: The Turning Groove resulting operation

We should have something similar to what is shown in *Figure 5.32*. Please note that the color inside the groove is green, meaning that the result is correct.

It is now time to move to the last turning command, which is very similar to grooving: **Turning Part** (also referred to as **Part Cutting** or **Parting**).

Turning Part

Turning Part is a command that allows us to detach our part from the stock cutting it. As you may imagine, cutting a part from the stock is typically the final operation to perform when turning a part. This type of operation is somewhat similar to grooving, with two major differences:

- When cutting, the cut width is not driven by the geometry of the part. We can cut whatever width we prefer. Of course, a higher cutting width means more material wasted and higher cutting power needed. Whereas, on the other hand, a super slim and super sharp cutting tool will cost more money and will wear in a shorter amount of time. Therefore, we should always aim at a trade-off between the material and the tool cost.

- When parting, our tool needs to plunge inside the stock, up until the rotation axis; therefore, cutting tools are usually longer and slimmer.

For the following example, we will need these two links:

- This is the link for the shank: `https://www.sandvik.coromant.com/en-gb/products/pages/productdetails.aspx?c=QD-NF-0250-0003-CR%20%201125`

- This link is for the insert: `https://www.sandvik.coromant.com/it-it/products/pages/productdetails.aspx?c=qd-nn2f33-25a`

In the previous links, we can find all the details needed to create our tool from scratch. Unfortunately, once again, Fusion 360 support for cutting tools is limited; therefore, we have to create a fake new cutting tool with a similar shape. Something like this should be fine:

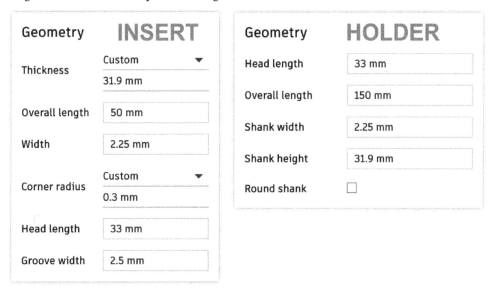

Figure 5.33: Custom cutting tool values

We cannot create a super fancy tool with a complex geometry, but as long as we specify the **Groove width** and the **Head length**, we can get a working result.

It is now time to launch **Turning Part** from the list of turning operations in *Figure 5.2*. As you may have noticed, the command is at the bottom of the list because, most of the time, it is the last command to set before the part is completed.

The Tool tab

Inside this panel, we can specify the tool we want to use and the main cutting parameters. For this cutting, we will use a cutting speed of **173 m/min**, a cutting feed of **0.14 mm** per revolution, and the maximum rotation speed supported by our chuck, which for my lathe is **4000 rpm**:

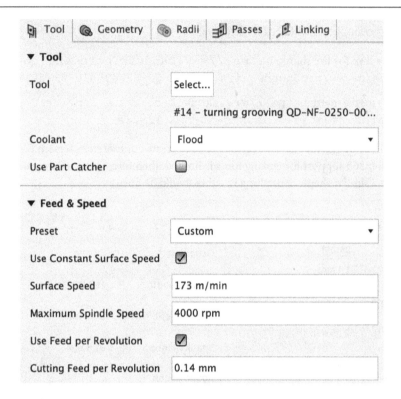

Figure 5.34: The Cutting Tool tab

Something to highlight here is an option called **Use Part Catcher**; as you can imagine, once the part is cut from the stock, it will fall to the ground. Some advanced lathes have a device called a **part catcher** that takes the machined part and puts it in a defined position. With this option, we can enable the use of such a mechanism (if both our lathe and our Post processor support it).

Another detail I want to highlight is **Maximum Spindle Speed**; as we already discovered in *Chapter 1*, when cutting close to the rotation axis, the cutting speed drops to almost zero. That's why, most of the time, when cutting a part, we have to set the maximum chuck rotation speed, in this case **4000 rpm**.

There is not much to say about the **Surface Speed** (set at **173 m/min**) and the **Cutting Feed** (set at **0.14 mm**), as both these values were calculated by CoroPlus, and simply copy and pasted.

The Geometry tab

Inside this panel, we can specify the position where we want to perform the part cutting, and whether we want a sharp edge, a chamfer, or a fillet on the edges:

Figure 5.35: The Cutting Geometry tab

Here, we can specify the plane where we want to perform our cut; most of the time, the default setting, **Model back**, is fine.

A much more interesting option is called **Edge Break**; with this option enabled, we can decide to create a chamfer or a fillet at the back of our part. This is something we definitely need for our example since there is still a chamfer we haven't dealt with yet:

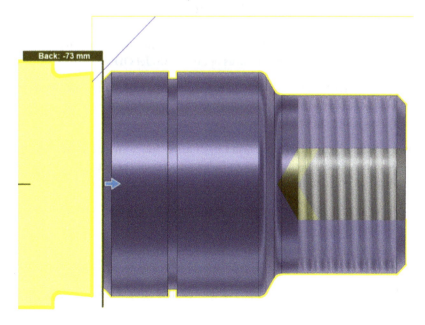

Figure 5.36: Back chamfer to be machined

As you can see from *Figure 5.36*, **Edge Break** created a 45° cutting path to machine the chamfer on the back of our part, and we will see the results in just a moment.

The Radii tab

Inside this tab, we can specify the working positions for our operation. It is very similar to what we have already found several times previously:

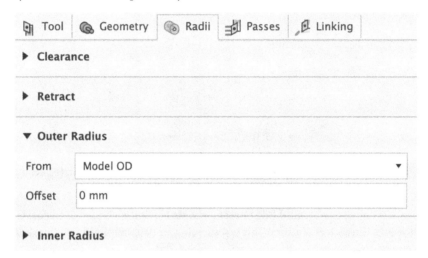

Figure 5.37: The Cutting Radii tab

There is not much to say about this tab, but I'd like to mention one important thing: for **Edge Break** to work properly, we need to set the **Outer Radius** option to **Model OD**.

As you may remember, **Outer Radius** is an option that allows us to trick Fusion 360 into thinking that the stock is a little bigger than the real one. We discovered that this could be used as a safety margin if we are not really sure about the stock diameter.

Setting a bigger diameter or an **Offset** value will force Fusion 360 to machine the chamfer starting from the wrong position; therefore, the chamfer won't be machined correctly. That's why we need to set the **Outer Radius** value to **Model OD**.

However, our part has already been the subject of several machining passes; therefore, we should not expect the diameter to be bigger!

The Passes tab

The **Passes** tab should look quite familiar to us, and indeed, it is quite similar to other operations we have already discovered:

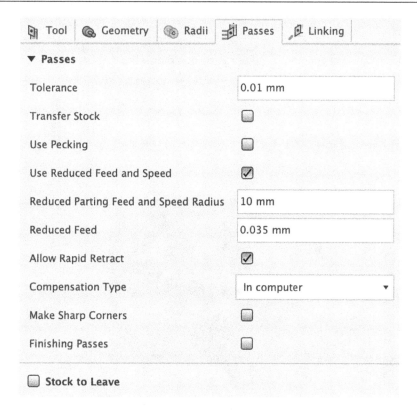

Figure 5.38: The Cutting Passes tab

There are a bunch of interesting options to analyze in this tab:

- **Transfer Stock**: If our lathe has two chucks, we can grab the part so it won't fall once the cut is completed. An additional chuck can be used for further machining or as a part catcher.

- **Use Reduced Feed and Speed**: As previously mentioned when discussing grooving, with this option, we can reduce the cutting load once the stock diameter left to cut is quite slim.

- **Reduced Parting Feed and Speed Radius**: This is the diameter, after which cutting speed and feed will be reduced to the new value.

- **Reduced Feed**: This is the feed value that will be assumed once the tool is cutting below the specified radius.

Simulation

We have completed our final operation, and our part should now be completed. Therefore, it is now time to check the simulation results to see whether everything was set up as we wanted:

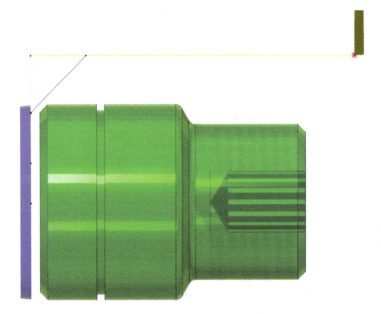

Figure 5.39: Turning Part simulation results

I think we should be more than happy with the result; as you can see, the part is detached from the stock and the chamfer on the rear was correctly machined. What a great piece of machining!

Summary

Congratulations! You completed the section related to turning. As you may already know, we didn't really cover every single command or option available, but we can definitely say that you now know a major part of the CAM module.

As a quick recap, first, we learned how to set up roughing passes to shape our geometry with longitudinal operations. Next, we discovered how to set finishing passes with a contouring strategy.

After that, we discovered how to create a thread and how to drill a hole inside the part. And finally, we reviewed how to create a groove and how to cut the part. With all these operations in mind, we should be ready to set toolpaths for almost every shape machinable with our lathe.

It is now time to move forward from turning and start talking about milling. As we will discover, most of the key concepts will be almost identical, so we can say that the effort spent up until now will earn you a lot of additional knowledge!

Part 2 – Milling with Fusion 360

In this part, we will leave turning behind and move on to a more advanced machining technique – milling. Once again, we will build our knowledge of the topic starting from scratch; we will review all the most important milling strategies and their potential, to prepare ourselves to move on to creating a rather complex component.

In addition, to further improve our expertise on the topic, we will analyze the common mistakes encountered when dealing with milling geometries so that we know how to avoid them.

This part includes the following chapters:

- *Chapter 6, Getting Started with Milling and Its Tools*
- *Chapter 7, Optimizing the Shape of Milled Parts to Avoid Design Flaws*
- *Chapter 8, Part Handling and Part Setup for Milling*
- *Chapter 9, Implementing Our First Milling Operations*
- *Chapter 10, Machining the Second Placement*

Getting Started with Milling and Its Tools

In this chapter, we will start moving from the turning world (which we should now be a bit familiar with) to the milling world, which has lots of potential and wide horizons for us to explore. Of course, this will lead us to new challenges and additional complexities that we will try to overcome during this chapter and the ones that follow.

We will discover the difference between a three-axis milling machine and a five-axis one, review the most common milling operations, and take a glimpse at the weirdly shaped tools that are involved in the cutting processes.

As always, we will learn useful formulas for parameter evaluation and a couple of golden rules that may save your day in the future.

In this chapter, we will cover the following topics:

- Understanding what milling is and how it works
- Understanding the main cutting parameters
- Introducing the most common milling operations

Technical requirements

We won't dig too much into milling theory and best practices since this book targets beginners, so there aren't any mandatory requirements for this chapter. However, reading the chapters on turning theory, particularly *Chapter 1*, will enhance your understanding of the subject since most of the formulas and concepts involved are very similar.

Understanding what milling is and how it works

Milling is probably the most flexible and capable machining method, able to create the most complex geometries with good productivity at appealing (cheap and flexible) costs.

Contrary to turning, in milling, the spindle holds the cutting tool (and not the part) and it moves in multiple directions relative to the part being machined.

We will try to keep things plain and simple during this introduction, but there are so many things we must skip or oversimplify that purists may get a bit skeptical! Don't worry, though – the core concept we are about to discuss is based on a solid ground founded on theory (just consider that you may find several exceptions in the real world that we don't have time to discuss).

Let's introduce milling processes by splitting machining into two main categories according to the number of degrees of movement of the spindle relative to the stock:

- Cartesian machines (up to three axes)
- Multi-axis machines (more than three axes)

Let's explore these two families in more detail.

> **Note**
>
> Why do we always refer to movements relative to the part? The reason is very simple: there are so many different machine configurations. For example, we may have a gantry architecture with a spindle capable of moving along three directions (X, Y, and Z) and the stock standing still, or we may have a machine with a spindle that can only move along the vertical axis (Z) while the part is mounted on a working plate that moves along two axes (X and Y). These two architectures are very different, but if we think in terms of relative coordinates, they can be considered the same. For the sake of simplicity, from now on, we may omit the term "relative" if not strictly necessary.

Cartesian machines

Cartesian machines are by far the most common (and less expensive) option for a milling process. On this type of machine, the spindle moves (relative to the part) along the three axes of a cartesian coordinate system (*X*, *Y*, and *Z*).

By providing a set of three coordinates to the machine (via CAM), the cutting tool can reach any point inside the working area. The following figure depicts a very simple milling operation by a three-axis machine:

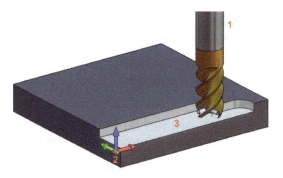

Figure 6.1: Cartesian machine

There are a couple of things worth noticing here. As you can see, the milling tool (*1*) is parallel to the *z* axis of the coordinate system (*2*), and it will maintain its orientation during the entire process. For a Cartesian machine, there is no way of tilting the tool in any direction in fact, so the geometry of machine (*3*) must be reached along the spindle axis only.

This is the main drawback of Cartesian milling machines – they lack a bit of flexibility when machining very complex shapes. Don't worry about this for now, though; we will expand on these types of problems in *Chapter 7* and *Chapter 8*.

Multi-axis machines

Machines with more than three degrees of movement are defined as multi-axis machines. These are milling machines where the spindle is not just capable of moving along the three main axes (like a Cartesian machine), but they can also tilt the spindle along one or more axis. The most typical milling machines on the market feature four or five axes.

The following figure shows what a five-axis machine is capable of:

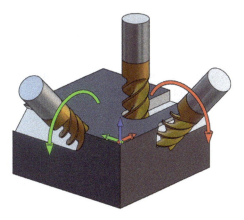

Figure 6.2: Five-axis machine

Not only is the tool capable of cutting the stock along the z axis (like a conventional Cartesian milling machine), but it can also rotate along the x axis and the y axis.

Thanks to modern electronic controls, this type of milling machine is gaining popularity at always cheaper prices. They also grant us high productivity and maximum flexibility when it comes to part geometry and complex shapes.

However, as a rule of thumb, we can say that multi-axis machines are still more expensive than three-axis milling machines and a bit more difficult to program via CAM for beginners.

Multi-axis milling centers give us great potential and a wide horizon of possible machining options, but they are not based on magic – they still have their limits. If not even a multi-axis solution can machine a part, we most likely have to redesign the part or move to another manufacturing process (something such as additive manufacturing, for example).

In the following chapters, we will create machining operations for a three-axis machine, but don't worry – these milling centers are still very powerful and capable (so capable that they sent a man to the moon!). Now, let's look at the main cutting parameters when it comes to milling.

Understanding the main cutting parameters

Milling is a complex process that requires us to understand many different parameters and working operations. However, we shouldn't be too scared of it since we should already be familiar with many turning operations.

In the following pages, we will discover how cutting parameters are conjugated for milling.

Spindle speed and cutting speed

Spindle speed and cutting speed are our old friends from turning operations; these concepts are the same for milling as well, so we will recap them a bit faster.

Spindle speed (n) is how fast our spindle (and our tool) rotates and is measured in **revolutions per minute (RPM)**.

Cutting speed (V_c) is the tangent velocity on the cutting edge measured in **meters per minute (m/min)** and can be calculated via the following formula:

$$V_c = \frac{\pi n D}{1000}$$

Here, D is the outer diameter of the cutting tool measured in millimeters.

The following figure shows a graphic visualization of the cutting and rotation speeds in the formula:

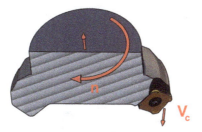

Figure 6.3: Cutting speed

Please note that, once again, the cutting speed is highly dependent on the radial position of the point of measure. Different points on the cutting edges have different cutting speeds. As a rule of thumb, we can decide to always use the outer diameter of the tool, but we should remember that there is a catch.

Cutting depths

As we should already know, the cutting depth is a parameter that measures how much the cutting tool is plunging inside the stock on a single cutting pass.

For a milling operation, there are two distinct parameters related to cutting depth:

- **Axial depth of cut (a_p)**
- **Radial depth of cut (a_e)**

Let's find out the differences between these two values by looking at the following figure:

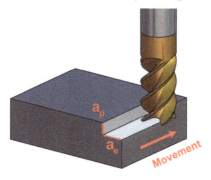

Figure 6.4: Cutting depths

As we can see, the tool is engaging the stock, milling a 90° profile on its corner. The height of the vertical face (parallel to the spindle rotation axis) is the axial depth of the cut (a_p), while the width of the horizontal face (perpendicular to the rotation axis) is the radial depth of the cut (a_e).

Both of these parameters are measured in millimeters and, as always, the higher the cutting depth, the higher the load on the tool.

It is important to note that we cannot set an arbitrary value to cutting depths. These two parameters are constrained by the geometry of the tool and the maximum cutting power available.

The following figure shows how to measure the maximum cutting depths on a tool:

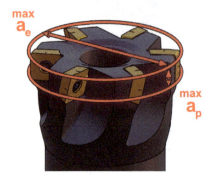

Figure 6.5: Maximum cutting depths

Notice that the maximum axial cutting depth is the height of the cutting edge, while the maximum radial cutting depth is the outer diameter of the tool. It is not possible to exceed those values on a single cutting pass.

Now, let's move on to the next parameter, which we should already be familiar with: the feed.

Feed step

The **feed step** is a parameter that measures how much the tool advances at every revolution of the spindle. The only minor difference between a turning feed and a milling feed is related to the different types of cutting tools.

As you may have already noticed, turning tools have one cutting edge only, while milling tools have multiple cutting edges. This is an important difference since all the cutting formulas aim to calculate the ideal workload on a single cutting edge.

At constant tool load, increasing the number of cutting edges (or inserts) will reduce the load on each cutting edge, while decreasing the quantity will increase the load. Therefore, in milling, we typically work with three different feed values:

- **Feed per tooth** (f_z)
- **Feed per revolution** (f_n)
- **Table feed** (V_f)

Let's take a look at each of these now.

Feed per tooth

Feed per tooth (f_z) is a parameter that measures the maximum thickness of the chip that every tooth is cutting, expressed in millimeters.

> **Note**
>
> In the mechanical world, the term tooth can be used to refer to gears and cutting tools. In these pages, tooth is intended to be a synonym for insert or cutting edge.

In the following figure, every tooth is highlighted with a different color and a number, and we can see rainbow strips where the tool is about to cut:

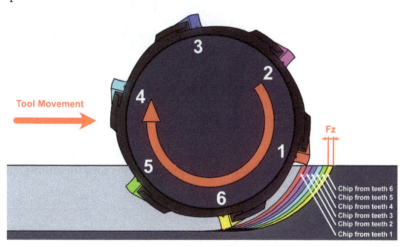

Figure 6.6: Feed per tooth

Every strip is the chip that is going to be generated by the corresponding teeth. The maximum thickness of the chip (along the cutting direction) is the value of the feed per tooth.

There are two main ways for a cutting edge to plunge into a part when machining, but the overall maximum chip thickness is not influenced by the approach of the cutting tool:

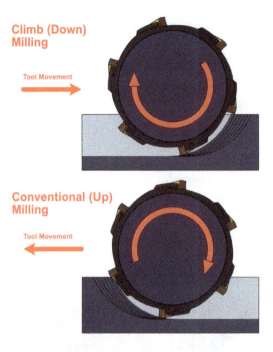

Figure 6.7: Up and down milling

As you can see, with **climb milling** (also called **down milling**), the cutting edge enters the stock where the chip is quite thick, while with **conventional milling** (also called **up milling**), the cutting edge enters the stock where the chip is thinnest.

> **Note**
>
> Climb milling is always preferred over conventional milling when possible. Peeling a very thin layer of chip from the stock is never the best way to go. The rule of thumb is to always enter the stock with a thick chip.

Feed per revolution

This parameter is equal to the turning feed that we use for turning operations. It is measured in **millimeters per revolution** (**mm/rev**) and expresses how much our tool moves along the cutting direction after a complete rotation of the spindle.

We can calculate the feed per revolution (f_n) via the following equation:

$$f_n = f_z * z_c$$

Please note that z_c is the average number of teeth engaging the stock, not the total number of cutting teeth of the tool. Multiple things can change the number of engaged teeth; we are about to discover a couple of elements to consider.

The following figure provides a visual representation of the number of engaged teeth:

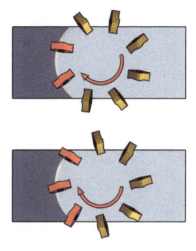

Figure 6.8: Engaged teeth

Here, the cutting tool is at an angle that engages two cutting edges in the stock. An instant later, since the tool is spinning, three teeth are engaged!

Another thing that changes the number of engaged teeth is geometry changes:

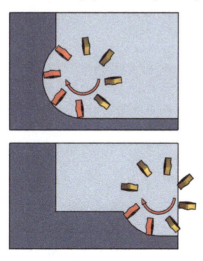

Figure 6.9: The engaged teeth change due to the geometry

As you can see, close to a corner, the number of engaged teeth can drastically change from the average number of engaged cutting teeth when machining a straight line – we pass from an average value of two teeth to four teeth engaged!

Changing the number of engaged cutting edges can modify the workload on the tool and lead to vibrations and resonances. That is why it is important to calculate the average number of engaged teeth.

However, thanks to modern CAM software, we can let the program create a toolpath with almost constant tool engagement.

If you want to explore a bit more about how a bad tool workload can ruin our parts, I strongly suggest you take a look at the following video by Haas Automation: `https://www.youtube.com/watch?v=rKPxfzx3sxE`.

Now that we have a general idea of what the milling feed is all about, we can move on to the next parameter, which is calculated using the values we just reviewed.

Table feed

Despite its name, the table feed (V_f) is not a feed value but rather the speed of our tool when cutting the stock. It can be calculated by combining the other values with the following formula:

$$V_f = f_z * z_c * n$$

Or, as an equivalent alternative, we can use the following one:

$$V_f = f_n * n$$

> **Note**
>
> f_n is measured in millimeters per revolution, while n is expressed in revolutions per minute; therefore, V_f is expressed in millimeters per minute.

As you may imagine, it is really important to always be sure to mention units of measure when dealing with feed values with other people. Confusing these three feed values in the formulas will lead to unforeseen effects on the part or our milling machine.

Cutting power and torque

Now that we have reviewed the main parameters involved in milling theory, we can approach the most important formula to evaluate the required cutting power (P_c):

$$P_c = \frac{a_p * a_e * V_f * K_c}{60 * 10^6}$$

Let's recap the elements we can spot in the equation:

- P_c is the cutting power and is expressed in **kilowatts (kW)**.

- a_p is the axial cutting depth and is expressed in **millimeters (mm)**.

- a_e is the radial cutting depth and is measured in **millimeters (mm)**.

- V_f is the table feed value; it is expressed in **millimeters per minute (mm/min)** and it is obtained by multiplying spindle speed (n) and feed per revolution (f_n).

- K_c is a parameter related to the material we are cutting and is expressed in **megapascal (Mpa)**, which you learned about in *Chapter 1*. However, since K_c is highly dependent on the feed step, we can no longer retrieve the same value of K_c from the table we used for turning operations. The table we used in turning was expressed as a function of the feed step; however, as we just discovered, in milling, there are multiple types of feed values. To calculate K_c in milling, we have to work with feed per tooth. Luckily, we can find many charts and tables for a proper approximation of K_c, such as http://www.mitsubishicarbide.com/en/technical_information/tec_rotating_tools/face_mills/tec_milling_formula/tec_milling_power_formula.

The previous formula is very useful since not only does it let us calculate the cutting power required by a certain operation, but also if we flip it, we can use it to calculate one parameter if we already know the others.

For example, let's suppose that we are wondering what the best axial depth of cut is for a certain operation. In that case, we would know the cutting power of our milling machine. At a given table feed, V_f, we know the feed per tooth, so by using a chart or a table, we can calculate the K_c value with ease. So, if we know all the other values in the formula, we can calculate the axial depth of the cut as well.

Now, moving on, to evaluate the torque, which is expressed by **Newton meter (Nm)**, we can use the following formula instead:

$$M_c = \frac{P_c * 9549}{n}$$

The formulas we've just discovered can be used both to calculate a cutting parameter we don't yet know and to check if an operation we are planning to perform can be sustained by our machine. Depending on the maximum power and the torque of our spindle, we may need to reduce the workload if the required settings are too demanding.

Please note that our setup must verify both cutting power and torque requirements.

If all of those formulas sound a bit tricky to you, don't be scared – multiple online calculators can do the math for you. CoroPlus is another great tool that can give us valuable suggestions on how to set all those parameters.

For now, let's move on to the next section, where we will discover the main milling operations.

Introducing the most common milling operations

Milling flexibility translates to a wide range of different machining operations that can realize almost any shape conceivable.

We can divide milling operations into five main categories according to the tool used and the machined shape:

- Face milling
- Shoulder milling
- Slot milling
- Profile milling
- Other

Let's introduce them one by one in the simplest way possible.

Face milling

Face milling is one of the most common and most simple milling operations. Its goal is to machine a flat surface perpendicular to the tool's rotation axis. It can be considered both a roughing operation and a finishing operation according to the tool used and its cutting parameters. The following figure shows a typical facing operation:

Figure 6.10: Face milling

As we can see, face milling tools can have complex geometries; they often have several replaceable inserts with the strangest shapes. These tools can be quite big and heavy and according to their dimensions, they can also be quite expensive.

A useful way to classify a facing tool is by using its entering angle:

Figure 6.11: Different entering angles

The most direct consequence of different entering angles is the direction along which the cutting force is exchanged between the cutting tool and the machined stock.

On the left of the preceding figure, there is a tool with a 15° entering angle; with such an entering angle, the cutting force is mainly transferred along the axial direction of the tool. Such a small entering angle reduces chip thickness and therefore allows a much higher feed step.

Facing operations with this type of tool have high productivity at low vibrations (since the radial force component is almost zero). Also, note that due to the insert's small angle, the maximum cutting depth is quite shallow.

On the right of the preceding figure, there is a tool with a 90° entering angle. The cutting force of this type of tool is transmitted as a radial component only, so it is more prone to runout vibrations – they are not optimal from the point of view of chip formation either, so they lead to overall lower productivity.

However, operations with a 90° entering angle allow us to have a higher depth of cut and are useful for most mechanical components where there are many perpendicular surfaces. Also, a tool with such an entering angle can be used for other types of milling operations as well, so it is flexible and cost-saving (since a single-facing tool with a diameter of 100 mm may cost more than 500 euros).

In the center of the preceding figure, we can see a tool with the most common entering angle for facing operations. It is a tool with an entering angle of 45° and its characteristics are in a sort of sweet spot between the other two.

Now that we have discovered the very first milling operation, we can jump to the next one, which is in many ways similar to face milling.

Shoulder milling

Shoulder milling is quite similar to face milling, with the notable difference that the main machined surface is not perpendicular to the spindle axis but rather parallel to it. Once again, shoulder milling can be used as a roughing operation, as well as a finishing operation. There are multiple tools capable of performing shoulder milling, but for the sake of simplicity, we will stick to end mills only:

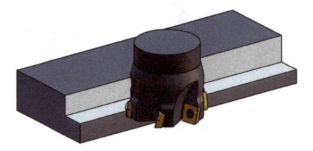

Figure 6.12: Shoulder milling

As we can see, sometimes, it can be difficult to understand if we are looking at a facing operation or a shouldering operation; there is not a clear and easy-to-understand distinction between face milling and shoulder milling.

For example, a face milling operation that employs a tool with an entering angle of 90° is performing a face milling operation on the bottom face of the cut and a shoulder milling operation on the sides at the same time.

Having already introduced a couple of milling operations, I'd like to spend a few moments on two golden rules to always apply in milling (I promise that this extra effort will be worth it).

Choose the right milling diameter

One of the golden rules in milling (especially for roughing operations) is to always use the biggest tool possible. The bigger the diameter, the faster the operation and also the better the surface finish due to fewer vibrations at higher workloads.

As you can imagine, due to geometries or cutting power restrictions, sometimes, it is not possible to pick the biggest and toughest tool on the shelf. As a rule of thumb, to pick the right diameter for our operation, we should always select a tool whose diameter is at least 30% or 40% bigger than the cutting width (also called radial cutting depth) we intend to set.

Mind the centerline

You should also remember to take care of the centerline placement. This may be a bit more complex to understand as it's all about cutter diameter and radial depth of cut, but check out the following figure:

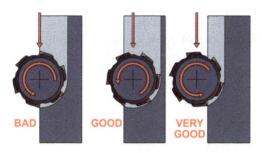

Figure 6.13: Centerline placement

As you can see, on the left, we have a tool cutting at a very particular cutting position; the depth of the cut (a_e) is equal to half the diameter of the cutting tool. This leads to having the centerline of the tool aligned with the point where the cutting edge enters the stock. It is a less-than-ideal condition since it causes high vibrations and faster insert wear.

In the center of the figure, we can find a much better situation where the radial cutting depth (a_e) is around 75% of the tool diameter; this is an ideal position for a milling operation as it improves chip formation and reduces vibrations compared to the first example. There are very complex and painful formulas to demonstrate why this is a better approach than the previous case, you just have to trust me, unfortunately!

An even better scenario is shown on the right, where the radial cutting depth (a_e) is around 25% of the tool diameter. This way, we are also respecting the first golden rule since the cutting diameter is way more than 30% bigger than the radial cutting depth!

Slot milling

Slot milling is a milling operation where the tool plunges inside the stock and machines the material into a shape, similar to the keyway of a shaft. Several slot milling techniques are using different tools but we will only analyze those machined with end mills since they are the most flexible.

The following figure shows a closed slot with quite a complex profile:

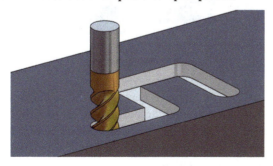

Figure 6.14: Slot milling

At first glance, we may think that we are not dealing with something tricky, but in reality, this type of operation can be really painful for our tool! As we can see, machining slots with an end mill allows us to machine almost any slot path (which is good news). However, creating a slot (especially if closed) with an endmill is a demanding task for the following reasons:

- If the slot is closed, our tool is forced to plunge into the stock axially, which is never a good way to approach the part

- Most of the cutting edges are constantly engaged

- For deep slots, chip evacuation is a problem we have to seriously consider

- Our tool is forced to operate in conventional milling, and as already mentioned, up milling is not the most ideal approach for a milling operation

- According to the slot shape, the number of engaged teeth may instantly change leading to strong vibrations

Wow, that was unexpected; there are so many complexities for such an apparently simple operation. Ignoring hidden troubles when approaching CAM may cause lots of painful frustration.

To simplify your experience with slot milling, I'd like to give you one useful tip just in case you need to machine a proper shaft keyway with a tight tolerance one day. As shown in the following figure, the first shaft is machined with an endmill whose cutting diameter fits perfectly the slot. The cutting toolpath is moving the tool back and forth; this is never the best solution as it will cause a bad surface finish and the slot walls may not be perpendicular to the bottom of the slot due to the high workload:

Figure 6.15: Slot milling tip

A much better approach to proper slot milling is depicted in the second example, where the slot is milled with a smaller tool that moves around the slot contour. This option is better because most of the troubles that we learned are related to slot milling and will be present only during the first approach to the stock; after this, the tool will machine the rest of the material as a typical shoulder milling performed in a climb cutting operation.

Profile milling

Profile milling is an advanced milling operation that involves machining a complex smooth surface along multiple directions. For this type of operation, a multi-axis machine shines, but even a standard Cartesian machine can express its potential if the geometry is not very complex.

Profiling is almost always used as a finishing operation; therefore, it is important to always prepare the surface to machine with previous roughing operations.

As shown in the following figure, the idea behind this machining technique is to let the tool follow the surface to machine with very small incremental steps; the smaller the step, the better the finish quality:

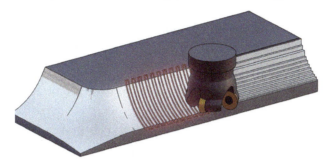

Figure 6.16: Profile milling

On the right, we can find multiple sharp edges (similar to a set of stairs) left from previous roughing operations yet to be machined, while on the left, we can find the shiny surface already finished.

For this operation, it is critical to reduce the vibrations to the bare minimum, so a short tool and a solid fixture are mandatory. The most common group of tools used for profiling include round inserts and ball nose end mills.

Please note that there are multiple profiling techniques according to the shape to be machined. Every profiling technique will create different toolpaths; however, the final result will be more or less similar.

Other

As you may have guessed, *Other* is not a type of milling operation, but rather it is a group of very specific operations that cannot be classified under the previous categories. There are so many additional milling processes we haven't mentioned yet, from gear milling to turn milling; unfortunately, we don't have time (nor space) to cover them all.

The only milling operation inside this group we will review is thread milling. It is a process where we can create a thread without using a tap or a die, much like we did when threading the example part on turning in *Chapter 5*. Instead, thanks to modern CNC machines, it is now possible to machine a thread with a special tool by applying a helical movement:

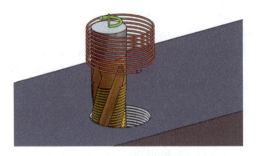

Figure 6.17: Thread milling

As we can see, the tool is performing a downward spiral path while simultaneously controlling the tool rotation, resulting in a threaded hole. There are several advantages to this technique instead of following a typical approach with a tap or a die:

- A single tool allows us to thread a wide range of threaded holes, while with a tap, we need one for each different thread dimension

- Chip evacuation is much easier since the tool is smaller than the hole

- It is possible to thread almost the entire length of a blind hole

We will try to implement all these operations using Fusion 360 in the examples provided in *Chapter 9*, and *Chapter 10*; we just need a bit more knowledge before jumping right into the action!

Summary

With that, we have introduced milling operations. It was not as detailed as it could have been, but a more complete overview would have required a dedicated book on the subject. I hope that the information provided has encouraged you to explore this amazing technology further.

Now, let's recap what we have learned. First, we understood how a milling tool can move relative to the stock, and the differences in degrees of movement between a traditional three-axis machine and a multi-axis machine.

Then, we plunged into the theory behind milling operations and looked at the main cutting parameters present in formulas for cutting power and cutting torque. After presenting those formulas, we learned how to use them to calculate a missing parameter when we know all of the others.

Moving on, we presented the main milling operations, and we provided a couple of useful suggestions on how to avoid hidden troubles when implementing our CAM operations.

Understanding all these concepts is the starting point for future chapters, where we will experiment with milling in a real-world scenario.

In the next chapter, we will discuss possible design flaws in milled parts and how to optimize the shape of our parts.

7

Optimizing the Shape of Milled Parts to Avoid Design Flaws

Sometimes, when working on a concept project, we can create our shape driven by its intended use or interactions with other mechanical components.

A usage-driven approach is perfect for achieving the most optimized geometry that will work the best. However, we may miss a few important details needed from a manufacturing point of view. These "minor details", however, may be the key difference between a cheap component and a much more expensive one or even an impossible part to machine.

In this chapter, we will explain the most typical issues with milled parts, showing you a wide set of tricky geometries to be milled and the approach to achieving them successfully. This should give you the tools for a more aware and better approach to part design.

In this chapter, we will cover the following topics:

- Handling undercuts and accessibility
- Learning how to manage mill radius
- Solving a very bad design

Technical requirements

The main technical requirement for this chapter is having a basic knowledge of part design using any CAD software. Making sure you have read *Chapter 6* is a great starting point since we will now dive deeper into milling, but this time from a design perspective. We will also discover a couple of commands from Fusion 360 that can help us during the design process; therefore, having access to the program is also recommended.

Handling undercuts and accessibility

There is a typical issue with milled parts that can be found in models designed without a clear understanding of manufacturing limitations: geometry **undercuts**.

What is this issue all about? The important thing to understand is that our machine can reach any point located inside its working boundaries, but the geometry we are trying to machine may restrict accessibility and impede tool movement. Areas not accessible to our tool are called undercuts.

The following diagram is an example of a badly designed part featuring an undercut:

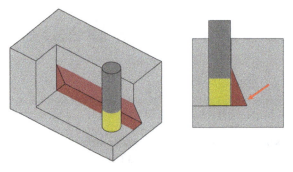

Figure 7.1: Undercut

As you can see from the diagram, the highlighted area is inaccessible to our tool; therefore, it cannot be machined.

Before starting any milling setup, we should always analyze in detail part geometry and check for undercuts. If the shape to be machined is quite complex, there is a useful tool inside Fusion 360 that can spot undercuts. This command is called **Accessibility Analysis** and can be found in the following menu:

Figure 7.2: Accessibility Analysis

This command is very similar to **Draft Analysis** you may have already used inside the **DESIGN** environment of Fusion 360 or any other CAD software.

> **Note**
>
> Draft analysis is a geometrical analysis of the orientation of surface normals. This analysis is typically used for mold design to check whether an object can be removed from the mold once solidified.

Accessibility Analysis is a simpler type of analysis, giving us information about positive or negative drafting angles:

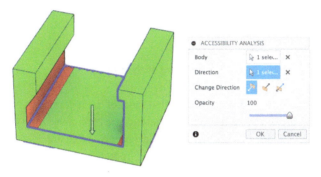

Figure 7.3: Undercuts spotted

As you can see from the screenshot, we can select a **Body** to analyze and a **Direction** by simply clicking on a face or an edge. The surfaces highlighted in green are reachable by a tool along the specified direction, while red areas are to be considered undercuts.

Undercuts should always be avoided when designing a new part; however, sometimes, the component we need to machine must have an undercut geometry due to its mechanical function. If this is the case, we may need to try a few workarounds to machine undercuts successfully; however, don't get too excited, as we can only successfully solve minor undercuts. If the part was not really designed with milling processes in mind, it might be necessary to drastically change its shape. Let's examine a few typical solutions.

Changing part orientation relative to the tool

Sometimes an area of our part may be unreachable for the tool if oriented along a certain direction, but it may become machinable if we change its orientation.

Figure 7.4: Undercut analysis profile

As we can see here, the accessibility analysis shows that most of the object can be machined along the z axis, while a few areas are not reachable. These undercuts are reachable along the x axis; therefore, we can set up two placements for machining. Multi placements are an advanced type of machining where the stock is moved or rotated between one cutting operation and the next.

For example, we can first decide to machine the slot along the x axis and then to machine everything else along the z axis:

Figure 7.5: The x axis direction

After fixing the stock as shown, the slot can be machined without undercut-related issues. After this operation (the first placement) we can open our CNC machine, remove the partially machined stock from the vise that holds it in place, rotate the part, close the vise again, and machine all the other faces (this would be the second placement):

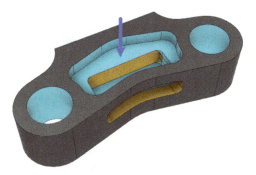

Figure 7.6: The z axis direction

Don't worry about the machined leftover radii that are left on the part because we will dive into them in the following *Learning how to manage mill radius* section. Part orientation and the order of machining placements are difficult subjects to master; there is not one general solution that works in every scenario.

Also, sometimes the final scope of the component we are machining may force us to use a less-than-optimal approach from a manufacturing point of view. It is always a good idea to keep a constructive dialogue open with parts designers to understand whether certain problematic features on the part are strictly needed or whether it is possible to tweak the shape a little to optimize production times and costs.

Using a multi-axis machine

Sometimes even changing the part orientation is not enough to solve accessibility issues because the features to be machined smoothly change their normal direction. For example, in the following diagram, there is a very complex slot profile to be machined on the cylindrical surface:

Figure 7.7: A complex profile

Working on this part with a basic three-axis milling machine is a problem. In order to achieve the feature as shown, we would need to change part orientation an infinite amount of times.

In this type of scenario, such a part requires the use of a multi-axis machine capable of tilting the spindle (or the part) along multiple rotation axes. Please note that since five-axis machines have a much higher degree of complexity, we will stick to conventional machines with three axes only.

> **Note**
>
> A normal vector to a surface is a vector which is perpendicular to the surface. By convention, the tip of a normal vector to a surface points outwards. For example, a normal vector to the face of a cube is oriented inside out. Curved surfaces have a normal vector constantly changing on every point of the surface.

Creating custom tools

Sometimes, part of the geometry we have to machine is inaccessible even when using a multi-axis machine. This is the worst-case scenario, but we may still have a last ace up our sleeve.

As a last resort, we may decide to build a custom tool shaped in such a way as to reach the inaccessible area. This is a technique that is very often used for wood machining; it is not a great method for harder materials such as steel, but it is still possible in principle:

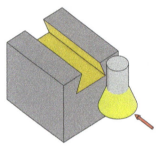

Figure 7.8: A custom tool

As you can see, it is impossible to machine such a sharp undercut using a typical cylindrical flute; however, with a properly shaped mill, we can machine very complex geometries with ease. The key idea of this type of operation is that the tool shape is imprinted in the stock while machining; therefore, a complex tool profile will lead to a complex part shape.

Fusion 360 features a very simple way to create custom tools; let's find out how with the following example part:

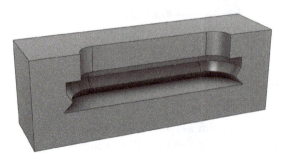

Figure 7.9: A complex profile

This part features quite a complex profile machined inside the stock; with so many sharp angles and low accessibility areas, there is no way we can machine this.

For this, we could build a custom tool that reflects the machined profile. In order to do this, we can create the profile for our tool with a simple 2D sketch inside the **DESIGN** environment of Fusion 360:

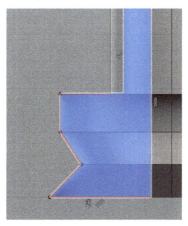

Figure 7.10: A custom profile

As you can see, I created a closed sketch that copies the profile of the part. Please note that you shall draw the shank profile and the spindle axis as well.

Once the sketch is completed, we can open the **MANUFACTURE** environment and launch the following command, **Form Mill**:

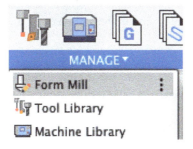

Figure 7.11: The Form Mill command

With **Form Mill,** we can create a custom tool providing a sketch as input:

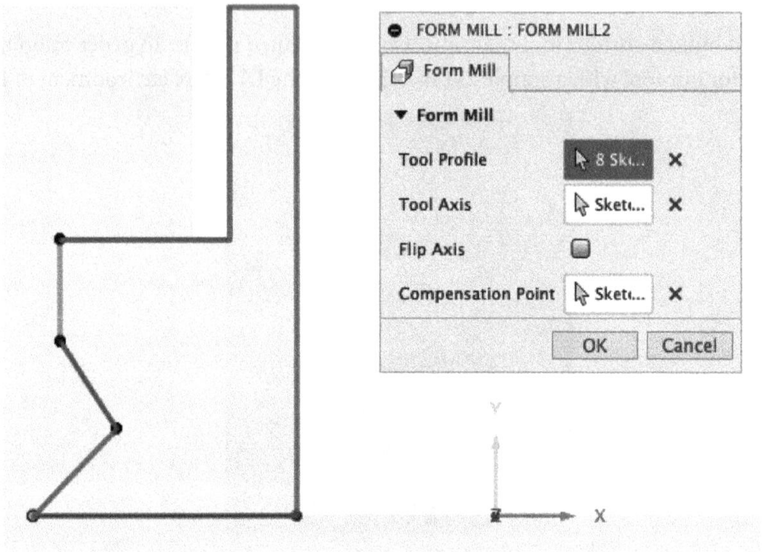

Figure 7.12: The Form Mill command

The interface is pretty straightforward but let's review it in detail:

- **Tool Profile**: With this selection box, we have to pick the closed sketch that we created earlier
- **Tool axis**: With this selection, we can specify the rotation axis of the spindle
- **Compensation Point**: This is the point that is going to be used to create toolpath coordinates; most likely, you want to set this point on the bottom of the tool or the corners of the profile

Once everything is set correctly, we should find the resulting tool inside the tool library, ready to be used!

Not only can this type of form tool be useful for solving undercuts issues, but it can also speed up the machining of complex profiles. A form tool can machine, with a single pass, what a cylindrical tool may machine in several operations.

Getting back to undercuts, if not even a custom mill is capable of machining our part, chances are that nothing else will. If this is the case, we should probably change the part shape or change the manufacturing technology and opt for something a little bit more flexible, such as additive manufacturing.

Backside milling

Backside milling is another advanced technique that can help us to reduce the total number of placements, solve undercuts issues, and improve tolerances.

It can be a bit tricky to understand how backside milling works, so think of it like this: imagine a conventional drilling operation, where the drill bit enters the part pushing against the stock; with backside milling, our tool enters the part pulling against the stock!

For example, let's suppose we have at our disposal a three-axis machine and that we want to machine both sides of the stock without setting multiple placements:

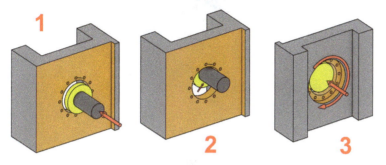

Figure 7.13: Pull milling

As you can see, on the right, we have the front face of our stock already machined successfully. However, we need to machine the back face as well (which is, of course, inaccessible from the front).

However, if there is sufficient room, with a special tool, we can move the tool through the hole (*1*), move it radially at a safe distance from the stock (*2*), and then start pulling the tool against the stock and machine the back face (*3*).

As you can imagine, this is a very tricky technique that can be implemented in only a few cases due to the limited access to the undercut area. However, it is worth mentioning since it may help us machine an otherwise inaccessible area!

I hope that exploring all these techniques to solve undercuts will help you in designing better parts and in solving accessibility issues. It is now time to move on to the next typical issue found on milled parts.

Learning how to manage mill radius

Having worked with milled parts for years, I can assure you that another common issue you will encounter is the wrong management of the mill radius.

To overcome this, you need to understand the following concept: milling processes cannot perform squared slots, and the mill radius will always be imprinted on the part.

In the following diagram, we can better understand the problem:

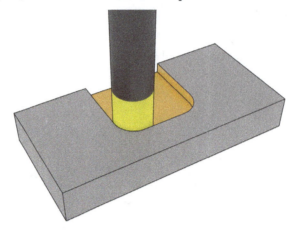

Figure 7.14: Mill radius

As you can see from the diagram, there is no way for our flute to leave a rectangular-shaped slot. If the mill cutter has a 20 mm diameter, we will always leave a 10 mm radius on corners. But what if we really need a squared slot? There are a few techniques that can help overcome this intrinsic behavior of milling machines:

- Reducing the mill radius
- Tweaking our part geometry
- Changing the milling direction

Let's look at these three options in more detail.

Reducing the mill radius

This is by far the simplest option we have, as it doesn't involve any changes to the geometry. We simply need to change our milling operation with a smaller cutting tool.

We should be very careful when choosing this radius management strategy since a smaller tool always has several drawbacks when compared to a bigger one: it will cut much slower, it will wear faster, and it won't be able to cut at high depths. Therefore, the final cost of the part will be higher!

Having said that, in the following diagram, we can visualize what a smaller tool radius will lead to:

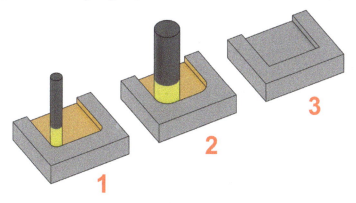

Figure 7.15: Mill radius reduction

Part *1* has been machined with a much smaller tool, and its angles feature a much smaller leftover radius than the ones on part *2*. However, we have improved the shape, but we still don't have the sharp angles shown in the designed part *3*!

Therefore, a smaller tool can leave a small enough radius that won't cause trouble for the components our part will be coupled with. And yet, sometimes, it still won't fit if the coupled part has sharp edges. Let's check out the following diagram for a better understanding:

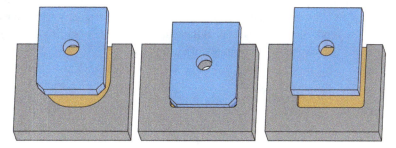

Figure 7.16: Mounting problems

Let's suppose we need to mount a part with chamfered edges inside the machined slot. The slot on the left has a leftover radius that is too big and will deny the coupling. In the center, we find a slot machined with a much smaller tool radius; it is smaller than the chamfered edge; therefore, we can couple the two parts together, which is great. However, on the right, we can see the worst-case scenario: the part we have to mount has sharp edges. In this case, no matter how small the tool is, the two won't fit, and in this case, we must choose another radius management option.

Tweaking our part geometry

The best option we have to allow a good coupling with a sharp part is a minor change to the geometry of our milled part. In the following diagram, we can view a typical example:

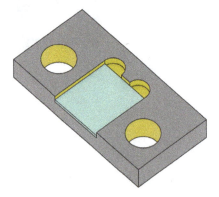

Figure 7.17: Shape tweak 1

As we can see, the part being mounted has sharp corners; therefore, it won't fit if there are any leftover radii on the corners of the slot. In the figure, however, it does fit because there are two bulges at the end of the slot. This way, we can have a perfect mechanical stop, and we can also machine the slot with a bigger tool!

This is the most flexible and advanced strategy, and it can be conjugated in a variety of different shapes. We just have to find the shape that best fits our scenario. For example, you may have noticed in *Figure 7.17* that there is not much material left at the end of the slot to work as a mechanical stop. A much better option is the following arrangement of the bulges:

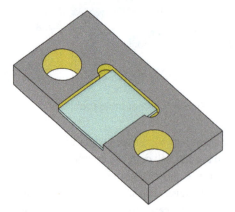

Figure 7.18: Shape Tweak 2

As you can see, we can still couple our part with a squared component, but we also have a much bigger area that works as a mechanical stop at the end of the slot!

There is one more thing that can manage the leftover radius on our part: changing the cutting orientation. Let's discover together how tool orientation influences the radii.

Changing the cutting direction

We just discovered that to manage the leftover radii of a milled part; we can reduce the tool diameter or tweak the geometry to create something close to a Mickey Mouse profile. However, as you can imagine, changing the spindle orientation will lead to a completely different radii pattern as well!

From the following diagram, we can now immediately spot a part that was designed without considering the tool radius (*1*):

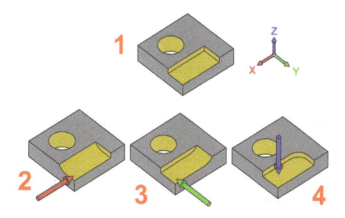

Figure 7.19: Spindle orientation

After understanding the intended function of the milled slot, we can decide the direction of our cutting tool relative to the part. For example, we can decide to orient the tool along the *x* axis (*2*), along the *y* axis (*3*), or again along the *z* axis (*4*).

Please note that these orientations will not only cause different radii placements but may also lead to different machining times and different final costs.

The best option to pick for this example is to only machine our part along the *z* axis; this is due to the fact that we must already machine the hole along the *z* axis. As you may have guessed, with a five-axis machine, we can smoothly change spindle orientation, while with a three-axis machine, we can change the cutting direction by rotating the part onto the working area with different fixtures.

Let's now suppose that, for whatever reason, we must machine the slot on our example part along the *y* axis with a three-axis machine:

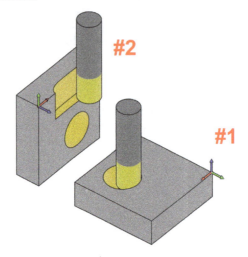

Figure 7.20: Machining a part with two placements

The first operation we can do is to fix our part so we can mill or drill the hole (*#1*); this is the first placement. After this first operation is completed, we can open the machine, remove the part from the first fixture, and mount it rotated (*#2*) onto a second fixture. Then, we close the machine and start a second machining task that will cut the slot.

This type of work can be done, but it takes more time, as well as requiring a trained operator who can mount the part correctly for the second placement without flipping it.

As you can imagine, it is quite more expensive than only machining everything along the *z* axis. Therefore, unless we have to use multiple placements, we should always avoid multiple spindle orientations as much as possible.

> **Note**
>
> You should definitely check out a very interesting video by Haas Automation to better understand this subject: `https://www.youtube.com/watch?v=oFvBe7cqxOE&list=WL&index=45`.

Getting used to spindle orientation, stock placements, and design errors are key points for advanced milling usage. In the following section, we will be presented with a 3D model with several machining issues. We will try to critically analyze what is and isn't feasible when machining the part.

Solving a very bad design

Now that we mentioned a few of the typical problems we may find with parts, we can review an example part to be machined from a squared stock. Let's suppose that we are CAM operators and that a group of designers give us this part to realize:

Figure 7.21: Example part to machine

This part is most likely an April fool's joke against us; there are a number of very big mistakes that not even a beginner can be excused for! But let's imagine this is the real deal and that we have to somehow machine it with our milling machine.

Let's review the mistakes one by one and try to find possible solutions with the design team.

Undercut face

There is a well-pronounced sloped face that cannot be machined at the moment. As we are about to discover, there are only three options available to us to achieve a similar shape, but all of them require minor changes to the design.

There is no correct one; picking one strategy over the other depends on the part usage and requirements.

Inclined tool

The first approach I would try to machine the undercut area requires a minor compromise to the design. I would propose machining the sloped face entering the stock with an inclined tool, which is very simple to achieve with a multi-axis milling machine (it may be a bit trickier on a three-axis machine, but still achievable):

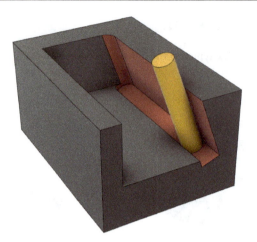

Figure 7.22: Inclined tool

As you can see, the sloped face is machined properly with only two minor drawbacks:

- There is a little bit of extra material removed on the bottom of the sloped face
- There is a leftover radius on the blind side of the slot.

If this radius compromises the function of the part, we could also opt for the following tweak in the geometry:

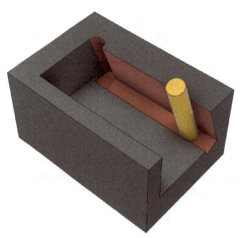

Figure 7.23: Radius management option

Now, the geometry is machined properly, and any part that may be installed inside the slot will still fit since we removed more material than intended in the original design.

Deep milling

If we don't have a multi-axis machine at our disposal, and we don't want to create a complex fixture to hold the part at weird angles, we could lay the stock down on one of the flat faces and machine it with a deep milling operation.

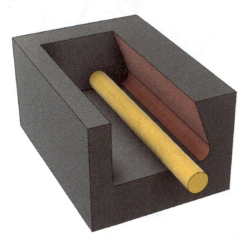

Figure 7.24: Deep milling

As you can see, entering the stock from the side will resolve the problem of the undercut and won't create a radius at the end of the slot. It could be a viable solution if we already have other features to be machined with the same tool orientation; however, in the example given, there aren't; therefore, it may not be the ideal solution.

In addition, we still have a leftover radius on the bottom of the slot, and we may need a very long tool to be able to machine the entire length. Using long and slim tools (like the one shown in the picture) is never a good idea, as they will bend more when machining; therefore, much lower feed steps must be used, and the total machining time will be affected dramatically.

Most likely, if the previous solution was not okay, this won't be either. This is definitely not the solution I would pick, but it is still worth mentioning.

Using a custom tool

The only other option, without a major redesign of the part, would be the usage of a custom tool. However, even with a custom tool, there will be leftover radii.

In the following example, with this type of tool, the bottom of the slot is flat, and the edges of the sloped face are sharp:

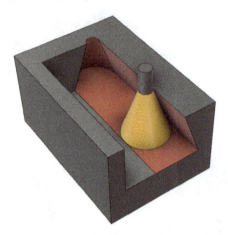

Figure 7.25: A custom tool

There are also a few drawbacks:

- On the end of the slot, there is a very pronounced leftover conical shape that was not featured at all in the designed part
- In addition, we should always keep in mind that custom tools are very expensive

For our example, I would say that, unless the sloped face features an angle of 60° or 45°, which is an angle we might already find in a tool catalog, I would not choose this type of machining.

Now that we have explored a few possible solutions to the sloped face – albeit solutions that won't completely work – let's choose one of them and move on to the next issue with the part.

Missing tool radii

A minor issue to fix is the lack of any consideration for tool radii; however, there is not much we can do on this topic. Due to accessibility issues, we can only choose to machine the part along the z axis or the y axis:

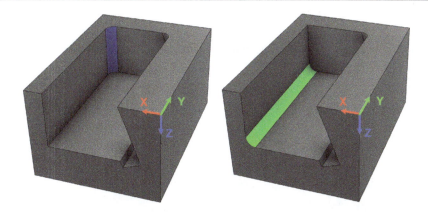

Figure 7.26: Machining direction

As you can see here, on the left, we have a leftover radius with a machining direction along the z axis, while on the right, it is set along the y axis. Choosing one or the other is dependent on the part's needs. However, if the choice doesn't compromise the part, we will opt for a milling process along the z axis.

The reason is quite simple: machining along y would require a much longer tool with all the drawbacks already discussed when evaluating deep milling.

We should now have clarified most of the milling process of the part, but there are two more issues to address.

Counterbore holes

The next major problem to fix is related to the counterbore holes on the sloped face we just machined. These holes are almost impossible to realize; the only option to have them as designed is to backmill them with an advanced tool, but it is something that is quite difficult, time consuming, and expensive to achieve.

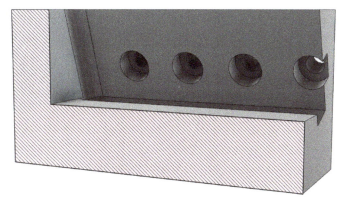

Figure 7.27: Counterbore holes

> **Note**
>
> You can take a look at reverse boring tools using the following link: `https://www.hermann-bilz.de/en/products/back-counterboring/`.

If we don't want to pick reverse counterboring tools, we can try a different approach to the part. For example, we could decide to reach the counterbore by extending it through the entire stock, as shown in the following diagram:

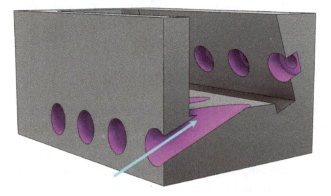

Figure 7.28: Back drilling

As you can see, if backside milling is not possible or desirable, we could decide to machine the counterbore faces from the back of the stock. With a long tool, we can machine the geometry, but also introduce holes on the bottom of the slot, which may not be good, depending on the part's usage.

For this example, I think the only realistic solution to the issue is investigating whether the counterbore faces really are needed and whether, with a minor design change, we can get rid of them because a much simpler option would be a simple hole all the way through.

As you can see, if the counterbore faces can be suppressed, we can access the part from the outside with the simplest drill bit, which would definitely be my choice if possible:

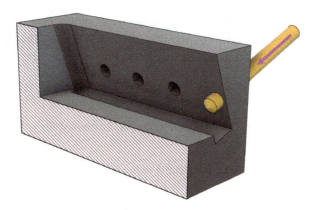

Figure 7.29: Simple holes

Now that we have discussed the hardest part to fix for this example, there is one last minor issue that would be nice to resolve.

All around radii

As you may have noticed in the part presented in *Figure 7.21*, all the exterior edges were smoothed out with a fillet.

Fillets are not impossible or difficult to machine; however, having them on every outer edge of the part would require us to machine all the fillets on one side, rotate the part for a second placement, and then machine all the fillets on the other side. Doing all these placements on a three-axis machine may be a little tricky; therefore, if all those fillets are not strictly needed, it would be preferable to remove those facing the bottom of the stock. This way, we can reduce the total number of placements, increasing productivity and reducing costs.

We just found a couple of tricks to approach complex geometries and a few workarounds to machine impossible shapes with minor compromises. I hope these suggestions help you in your everyday life as a CAM operator!

Summary

That was the end of the chapter, let's recap what we learned together. We introduced the basic concept of undercuts inside the geometry of the part and how undercuts are strictly related to tool orientation. Then we analyzed a couple of examples centered on how to machine undercuts if their presence is required by the part's intended use.

After this introduction to accessibility issues, we learned what leftover radii are and how to get rid of them if they compromise our part coupling with other components. Then we introduced the concept of multiple machining placements, and we found how they can change radii patterns and undercuts machining.

And finally, we tried to apply all the described techniques to a very tricky part to be machined.

This introduction to typical errors found on milled parts is essential for approaching the design of a part to be realized by milling operations.

In the following chapter, we will move on from the theory discussed till now, and we will start setting up milling operations with Fusion 360.

8

Part Handling and Part Setup for Milling

The most critical step when approaching milling processes is the setup. During the setup, several important decisions can greatly influence machining complexity, time, and costs even before turning the CNC on. For this reason, before jumping into action, we should always look closely at the geometry we need to machine, searching for hints about the best possible milling workflow.

In this chapter, we will try to analyze a complex part, starting right from the 2D drawing. We will learn how to extract important pieces of information from drawing details and how to lay down a possible workflow with multiple placements using WCS offsets. After this analysis, we will learn how to create a multiple-setup CAM project with Fusion 360.

The goal of this book, and this chapter in particular, is to change your point of view and move you from a design perspective to a manufacturing perspective. We will work on a real-life scenario, where we will be presented with multiple challenges and possible solutions.

In this chapter, we will cover the following topics:

- Understanding the part
- Choosing part placements
- Choosing a part fixture
- Choosing WCS offsets
- Defining the first setup
- Defining the second setup

Technical requirements

In order to follow along with this chapter, make sure that you have read *Chapter 6* and *Chapter 7*.

Understanding the part

The first thing we need to do before creating a proper CAM setup is focus on the shape of our part. We want to aim to have a complete understanding of the part's geometries, tolerances, and surface finish. Not concentrating on analyzing the part may lead to different machining approaches; that's why it is so important to take our time and check the drawings with the highest attention.

In the next few pages, we will explain how to machine the following part on a three-axis machine. The following figure shows a 2D drawing of the part we will be looking at:

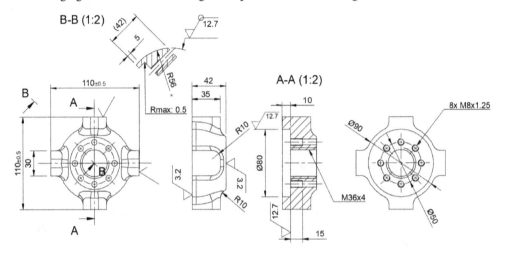

Figure 8.1: Part 2D drawing

What data can we extrapolate from this drawing?

- The component's overall dimensions are 110 x 110 x 42 mm.

- On the two planes defined by the 42 mm dimension, a good surface finish (3.2 microns) is required, so we should use a stock thicker than 42 mm and then machine it down to 42 mm with a finishing pass. For example, we can select a stock that's 45 mm thick and then machine it down to 42 mm.

- The two dimensions of 110 mm don't require a particular surface finish and they have a large tolerance either (±0.5 mm), so we may leave them unmachined. Leaving them unmachined means that we can use a stock of 110 x 110 x 45 mm. Of course, just in case we cannot find a raw material block with such a volume, we can pick a bigger one instead.

- There are features on both sides of the part, so we are forced to set two placements.

- There are several threaded holes, most of which are M8 x 1.25; these shouldn't be a problem with a simple drilling operation. In the center, there is a bigger one (M36 x 4) that we should probably machine with a different strategy.

- We have no information regarding the number of components we shall produce, but let's suppose a large number is needed.

- There's no mention of it in the drawing, but the material to be machined is a solid block of aluminum 6060.

Now that we have an idea of the shape we are about to machine, we have to understand what the best strategy for the part placement on the CNC working bed is. We already know that we need two placements, but which side of the stock should be machined first? Let's find out.

Choosing part placements

When studying part placements, most of the time, we can follow different paths, but some of them may be more complex than others. I'm going to explain what I think is the best solution regarding our specific example, but it may not be the only solution according to the type of machine at our disposal and the holding fixture we intend to use.

So, now that we have seen the part we plan to create in *Figure 8.1*, we have to imagine how to move from a stock of 110 x 110 x 45 mm to the final shape. As we discovered from the drawing, we have to machine both sides of the stock as the features will appear on both sides.

Since most milling operations have to be performed on the rounded side, I would machine this side last:

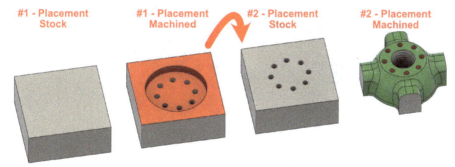

Figure 8.2: Part placements

As you can see in *Figure 8.2*, during the first placement (labeled as *#1*), I plan to perform the following operations:

- Back side facing

- Circular slot milling

- Drilling and threading the holes

We covered all of these in *Chapter 6*; however, in the following chapters, we will explore them deeply while setting up the operations via the CAM module.

After these operations, the first placement will be completely machined and the stock can be flipped and mounted upside down so that we can continue to machine the other side. The second placement (labeled as *#2*) contains all the major operations:

- Front side facing
- Circular hole milling and threading
- Milling of the round faces (we will have to implement a roughing and a finishing operation)

Please note that since the central threaded hole is accessible from both sides of the stock, we may decide to machine it during the first placement or the second placement, or both. Since thread milling can be quite complicated, I decided to leave it as the last operation (we will cover it in *Chapter 10*).

Now that we know what types of operations are involved, we can start thinking about the best way to hold the stock while machining it.

Choosing a part fixture

How do you fix a stock to the machine working area? There is no all-encompassing rule, but we can say that we should always try to take advantage of the part's shape to lock it onto the working area.

Let's review the part again to find some possible solutions:

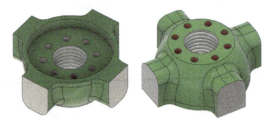

Figure 8.3: 3D view of the part

At first glance, a simple and cheap solution would be to use a stock vise to hold it from the squared unmachined faces.

As shown in *Figure 8.4*, using a large set of vises to hold the stock is a good solution during the first placement since there is a large contact area between the faces:

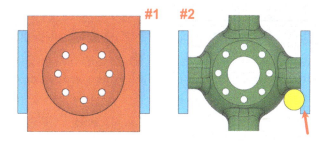

Figure 8.4: Part fixture

During the second placement, however, the same vises not only provide a smaller contact area that may lead to vibrations but they may also collide with the cutting tool performing the exterior machining.

Since we said that a large number of parts have to be machined, we may have to try a different solution – one that is a bit more complex. The solution I would pick is to realize a template to hold the stock in position.

> **Note**
>
> A machined template to hold the part in place can be quite expensive, sometimes even more than the part itself, so it becomes a viable option if we can charge its price on a large production batch of parts.

As shown in *Figure 8.5*, after the first placement, we machined a circular slot (Ø80) with several threaded holes. The idea is to take advantage of this feature to hold the part in place. To wrap things up, in the following figure, we can find a recap of how the operation is going to be performed:

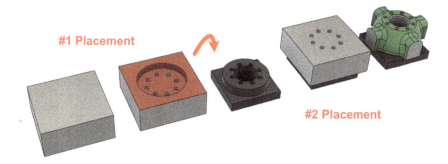

Figure 8.5: Advanced part fixture

The first placement (*#1*) is held in place with a typical set of vises. During the first placement, we machine a circular slot with several threaded holes.

To hold the stock in place during the second placement (#2), we decided to create a custom template (in the center of the figure) that will hold the stock in place while taking advantage of the threaded holes.

Please note that during the second placement, the stock is held from the bottom with screws so that there is complete access to the entire outer shape. This will allow us to use a bigger tool to improve machining time and costs.

Another typical way of reducing costs is to machine the two placements together. This way, at every machine stop, we have a complete part machined that we can pick from our CNC and put in a box and a new first placement stock that we can install onto the template for the second placement; then, we can repeat this cycle.

> **Note**
>
> We don't have the time to cover a large number of pages on this topic, but if you want to learn more about vises and holding fixtures, I suggest taking a look at this interesting video from NYC CNC: `https://www.youtube.com/watch?v=7og3lFitpSo`.

Now that we have an idea of how the stock is machined and held in place during the placements, we can go back to Fusion 360 to create the setup.

Choosing WCS offsets

As we mentioned previously, machining the two placements together can improve productivity and reduce overall costs. This takes us to an important choice we must define before any further consideration: do we want to perform a single setup for both placements or do we want to set two distinct setups each with its own coordinate system?

There is a minor difference between the two, but it is incredibly important to understand. Let's look at the following figure to help us:

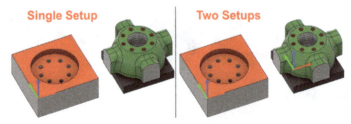

Figure 8.6: Single versus multiple setups

On the left, we can find a CAM program with a **Single Setup** for both placements. This means that the two placements are considered by Fusion 360 as a single part to be machined with the same origin for the **work coordinate system** (**WCS**). It is the simplest approach, but it has the dangerous drawback of forcing us to precisely fix the two components at the exact location specified inside the G-code program; any minor deviation will result in a badly machined geometry.

> **Note**
>
> What is a WCS? We already mentioned this topic but it might be worth highlighting it again: a WCS is any coordinate system used to measure the position of the tool. There is not a single WCS; we can set as many as we want. As you can guess, the same set of coordinates – let's say, (10;83;51) – applied to a different WCS will result in the tool moving to different locations.

On the right, there is a **Two Setups** program. As you can see, there are two disjointed origins with two different coordinate systems. It is the most flexible solution because it isn't as error sensitive as the previous case. Why is that? It's because we can take advantage of **WCS offsets**.

So, what are WCS offsets? Long story short, they are position bookmarks we can set for reference inside our machine. These bookmarks can be used to reset the tool coordinates to zero (basically, this means changing the WCS position).

If you have a nerdy background like me, think of WCS offsets as save slots of your favorite video game. Inside these save slots, you can store a certain position to be later used as a coordinate system.

Confused? I don't blame you; this is advanced stuff! I'll try to better explain how to use them in our example step by step:

1. We mount the stock to be machined anywhere onto the working bed using the vises. The machine has no idea of where the stock is placed, so we have to save the position of the stock origin.

2. We move the tool (or a measurement probe) until it is positioned exactly on the origin of the coordinate system we intend to use for the first placement. For example, if the origin is on one of the corners of the stock, we must move the tool until it touches that corner.

3. At this point, we can save this position as **WCS Offset number 1**. These coordinates will be stored inside the CNC memory. This way, once the CAM program needs to start from the origin of the first placement, the machine will know the exact starting point relative to the stock!

4. Now, we can run the CAM program for the first placement.

5. Then, we must flip the machined stock and mount it onto the template for the second placement.

6. Again, the machine doesn't know where the stock has been moved, so we have to let it know where the origin now is. Therefore, we must move the tool until it is positioned on the origin of the second placement.

7. This time, we must save this position as **WCS Offset number 2**.

8. Now, we can run the CAM program for the second placement.

Since we are using two setups, we can be sure that both zeroes will be maintained for both placements (doing this with a single setup would have been impossible).

> **Note**
>
> Since we have to manually set the zeroes for the two placements, it is a good idea to choose an origin located in a spot that's easy to measure. For example, in our case, I would never pick the center of the part; I would go for a sharp corner instead.

To wrap things up, you just have to remember that our CNC doesn't have eyes – somehow, we have to let it know where the stock and its origin are. Therefore, we have to save the position that our CAM toolpath will use as a coordinate system into its memory (WCS offsets). If we use two placements with one setup and one coordinate system each, we have to save two different WCS offsets.

Now that we understand WCS offsets and the potential of using two setups for a CAM program with multiple placements, we can open Fusion 360 and start using the CAM.

Defining the first setup

The manufacturing environment for milling is very similar to the one we explored for turning, and we should already be familiar with most of the panels and commands:

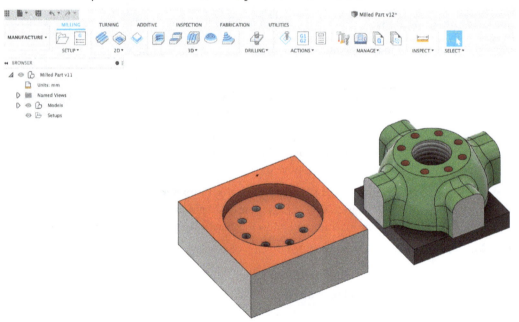

Figure 8.7: The two parts to model

As you can see, for better clarity, I decided to model two parts: the model on the left is a solid, showing how the part should look after the first placement has been completed, while the part on the right is the final result after the second placement has been fully worked.

Splitting the part into two distinct models – one for each placement – is not mandatory, but I think that considering the part as two distinct parts can be more intuitive for a beginner.

Now, let's launch the **Setup** tab. We should be presented with three panels:

- **Setup**
- **Stock**
- **Post Process**

Just like we have done in previous chapters, we will explore each panel one by one.

Setup tab

On the **Setup** tab, we can find the settings for fully defining the WCS placement relative to the model and the type of machining we are trying to implement:

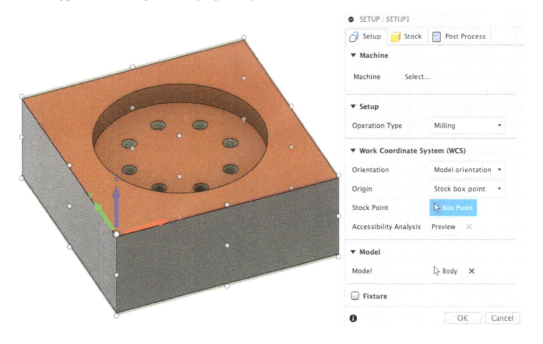

Figure 8.8: The Setup tab

Let's start with the most important setting.

First of all, we must be sure that **Operation Type** is set to **Milling** (and not, for example, **Turning**).

After that, since there are multiple bodies in the project, the next thing we must do is specify which 3D body has to be considered for the operation; the selection box labeled **Model** allows us to pick only the parts we want to machine. If nothing is picked, it will consider every 3D object as a part to be milled. In our example, we have the first placement stock, the final part, and the template, so we shall select the first placement solid only.

With the **Fixture** option, we can specify the vises that hold the part in place. However, since we are about to machine only the top side of the stock, and since my vises are not that tall, I can be pretty sure that no collision is going to occur, so we can leave this flag unticked.

Finally, we have to specify the **Origin** point for our WCS by selecting a point on the stock or the geometry. It is always a good idea to set a point that is easy to measure since we have to manually set the zero for our part.

Stock tab

On the **Stock** tab, we can specify the dimension of the aluminum block we have to machine:

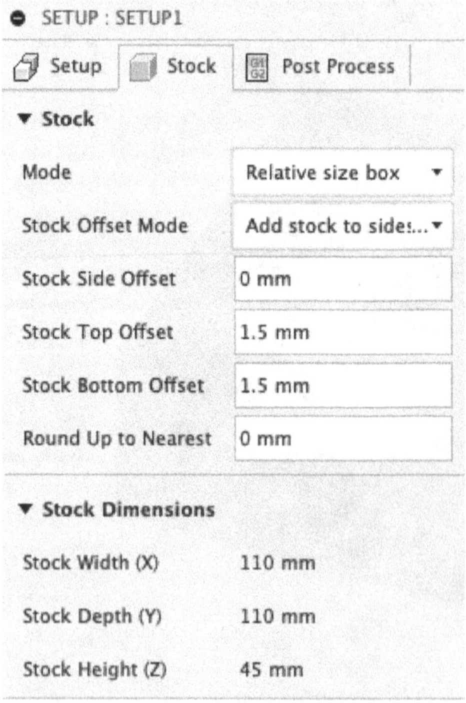

● SETUP : SETUP1		
Setup	Stock	Post Process

▼ **Stock**

Mode	Relative size box ▼
Stock Offset Mode	Add stock to sides... ▼
Stock Side Offset	0 mm
Stock Top Offset	1.5 mm
Stock Bottom Offset	1.5 mm
Round Up to Nearest	0 mm

▼ **Stock Dimensions**

Stock Width (X)	110 mm
Stock Depth (Y)	110 mm
Stock Height (Z)	45 mm

Figure 8.9: The Stock tab

The part we analyzed has a bounding volume of 110 x 110 x 42 mm; however, since the front face and the back face are required to have a good surface finish (and most likely a good planar tolerance), I plan to use a bigger stock with additional material along the Z axis. To ensure there's good planarity between the two faces, I would add at least 1.5 mm of material to both. Therefore, we should set a stock of at least 110 x 110 x 45 mm.

To set the stock bigger than needed, we can set **Mode** to **Relative size box**.

Choosing this option allows us to specify additional material relative to the overall dimensions of the part, instead of working with the absolute dimensions of the stock.

By clicking on the **Stock Offset Mode** dropdown, we can specify where we want these layers of material to be added and how thick they should be.

In this example, since we have to specify different thicknesses on the faces, we can pick **Add stock to sides, top-bottom** to customize the values independently.

As you can see, I left a **Stock Side Offset** of 0 mm since I don't plan to machine these faces at all, while I added 1.5 mm of material to both **Stock Top Offset** and **Stock Bottom Offset**.

To check that everything has been set properly, we can check **Stock Dimensions**. As intended, we discover a stock of 110 x 110 x 45 mm, which is what we wanted.

Post Process tab

There aren't many commands inside the **Post Process** tab, but here, we can find the option for the WCS offsets that we need to set:

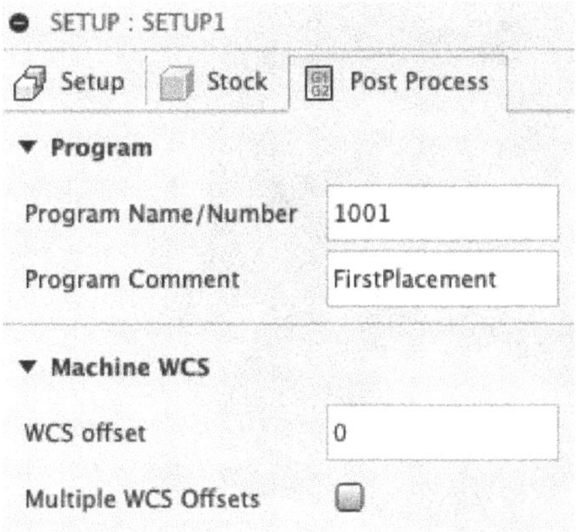

Figure 8.10: The Post Process tab

Be very careful! Inside this panel, we should be sure that **WCS offset** is set to **0** (which is the default value). You may be tempted to tick the **Multiple WCS Offsets** checkbox. However, *do not enable it*; this setup will only use one WCS!

> **Note**
>
> Please note that WCS offsets get different names during the translation between Fusion 360 and the G-code program. Inside Fusion 360, WCS offsets are referred to using plain numbers (0, 1, 2, and so on), while on almost any G-code flavor, they are converted into G-code commands (G54, G55, G56, and so on).

There is nothing else to change in this panel; therefore, we can consider our first setup completed. Now, we can set up the second placement.

Defining the second setup

In this section, we have to repeat more or less what we explored during the first setup. However, don't worry – we will only review the important differences from what we've already reviewed.

Setup tab

First, let's look at the **Setup** tab again:

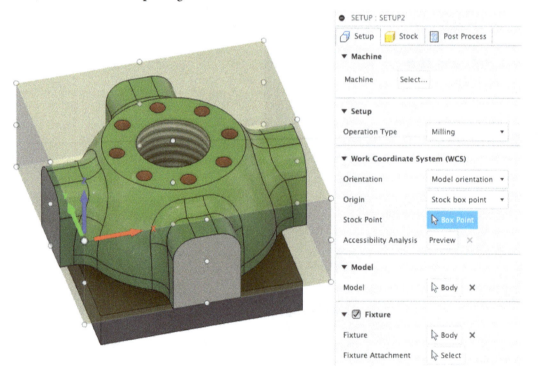

Figure 8.11: The Setup tab

As you can see, this time, I selected the final part to be machined, and I put the **Origin** parameter of the stock on the corner of the aluminum block.

This time, with the **Fixture** option, we can pick the template fixture that holds the stock in place. This way, our simulation will be able to check collisions between the tool and the holding fixture; this is not mandatory – if we did things properly, we shouldn't expect this type of issue – but we can never tell!

Stock tab

On the **Stock** tab, we only need to check if the overall dimension of the block is correct:

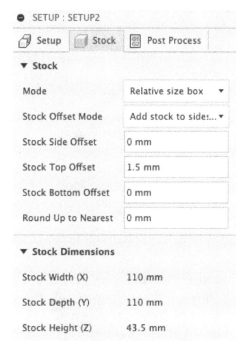

Figure 8.12: The Stock tab

As you may recall, the stock was 110 x 110 x 45 mm, but we plan to machine the back face during the first placement. Therefore, in this second setup, we should set a stock of 110 x 110 x 43.5 mm.

To set the stock dimensions with precision, we can use **Relative size box** once again. However, this time, we will set **Stock Bottom Offset** to 0 since we already machined that face in the previous setup. **Stock Top Offset** has to be 1.5 mm since we still need to machine the top face.

As always, we can double-check if the overall dimensions are as expected by looking under **Stock Dimensions**. As you will see, it will be 110 x 110 x 43.5 mm. Perfect!

Post Process

Now, let's move on to the **Post Process** tab:

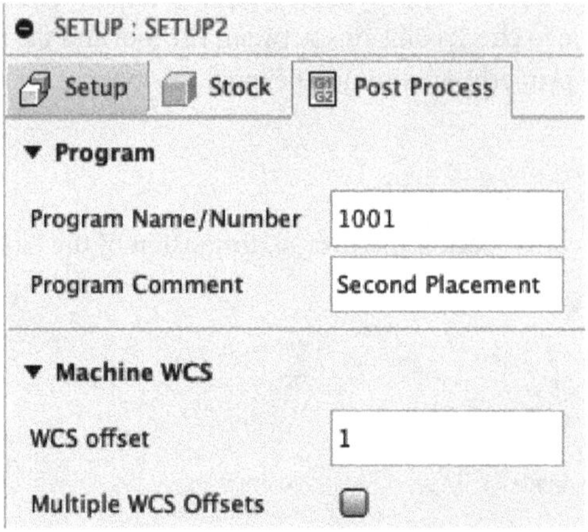

Figure 8.13: The Post Process tab

Here, the only value we should change is **WCS offset**. We have to change this number from **0** to any other value; for our example, we can use **1**. It doesn't matter what number is chosen (if it is supported by the post-processor); it will simply store the coordinate system in a different memory slot and won't impact the result.

> **Note**
>
> We don't have to insert G-code commands inside the WCS offset text box – we simply have to add a plain number. It's the post-processor's responsibility to convert that number into a G-code line.

Once again, be careful and do not tick the **Multiple WCS Offsets** option.

With that, we have completed the second setup! We should have something like this:

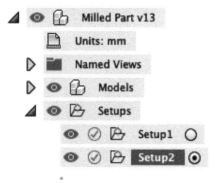

Figure 8.14: The Setups folder

As you can see, inside the **Setups** folder, we have two setups called **Setup1** and **Setup2**. Every milling operation for the first placement shall be saved inside **Setup1**, while every operation for the second placement shall be saved inside **Setup2**.

Now, we can start looking at milling operations.

Summary

That concludes this chapter! It may have seemed quite complex due to the introduction and usage of WCS offsets, but it is way simpler than it sounds; we will clarify this process in the next few chapters. Also, despite being a book oriented to CAM novices, the part we are trying to machine is quite complex and up until now has offered several examples of analysis that will come in handy in our daily machining.

To recap what we learned, first, we discovered the part we have to machine with our CNC. Then, using the drawing, we analyzed the part to understand the best setup strategy. After that, we analyzed the needed placements to be able to machine the entire geometry and discovered the best solution to hold the part during the second placement. Finally, we created two setups that will be used in the following chapters for our milling operations.

In *Chapter 9*, we will learn how to implement several milling strategies, such as face milling, shoulder milling, drilling, and tapping.

9

Implementing Our First Milling Operations

In this chapter, we will implement all the cutting operations to complete the first stock placement for our example part. We won't jump directly to complex 3D operations (which will be covered in *Chapter 10*) but rather, we will start from the most common 2D milling strategies of face milling and pocket milling.

We will learn how to set up the main 2D milling operations and how to calculate the cutting parameters using CoroPlus. While doing this, we will try to have a critical point of view on the proposed solution, learning how to optimize the machining time and reducing the number of tool changes in our CAM program.

Not only we will learn how to set up milling operations using an online calculator, but we will also calculate the cutting parameters by ourselves (step by step), using the equations introduced in *Chapter 6*.

The overall goal of this chapter is to get you used to the milling environment of Fusion 360 before moving on to more complex operations. This is very important, since jumping directly to fancy 3D contouring operations and finishing strategies may lead to confusion and frustration.

In this chapter, we will cover the following topics:

- Face milling
- Shoulder milling
- Drilling
- Tapping

Technical requirements

This chapter is a bit more advanced than the previous ones, so I strongly advise that you read and study all the chapters about milling and turning theory so far. Some concepts won't be explained again in the hope that you are already a bit more confident with the basics of CAM interfaces and machining operations.

Also, as we will be using CoroPlus for cutting parameters as well as solving equations, a calculator or an Excel sheet may be required.

Face milling

As described in *Chapter 8* (as shown in *Figure 8.2*), the first milling operation we have to implement is facing the top side of the stock; here, we have to remove a layer of 1.5 millimeters of our part.

At the moment, the stock is 45 mm tall, and we need to bring it down to 42 mm by removing 1.5 mm from both the top and bottom face, to have a good surface quality. As discussed in *Chapter 6*, the best strategy to remove a flat layer of material is face milling, a fundamental milling operation that allows us to machine flat areas.

> **Note**
>
> Depending on the quality of the stock, we may need to increase the amount of extra material to remove for a successful facing operation. The more the stock is bent or irregular, the more material we shall remove, and the bigger the stock shall be.

Let's begin with finding the right tool and parameters using our trusty Sandvik CoroPlus.

Using CoroPlus to find the best tool for face milling and shoulder milling

We have already used CoroPlus for turning; however, this is the first time we are approaching milling, therefore a fresh recap about this plugin may be useful.

The first thing we should set is the material of the stock, which in our case is an aluminum alloy series 6000. As you can see, we can choose between multiple default materials; every one has a color and a letter associated with it. We will select **N**, which in the Sandvik naming convention stands for aluminum alloys.

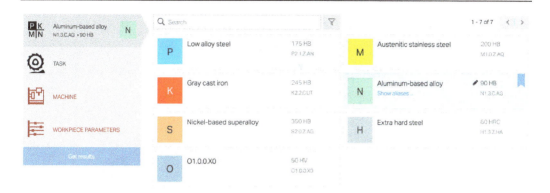

Figure 9.1: CoroPlus material selection

Picking the right material will ensure correct feeds and speeds, and the best possible optimization of the tool material and its coating. Please note that different brands will use different colors and naming conventions, so do make sure you are selecting the correct material.

Now, we have to select the right task to be implemented. This can be found in the same panel we already used for turning; however, this time, instead of picking **Symmetrical rotating**, we will go for **Non-rotating**.

Figure 9.2: CoroPlus task selection

Now, we will be able to see every milling operation we will ever need:

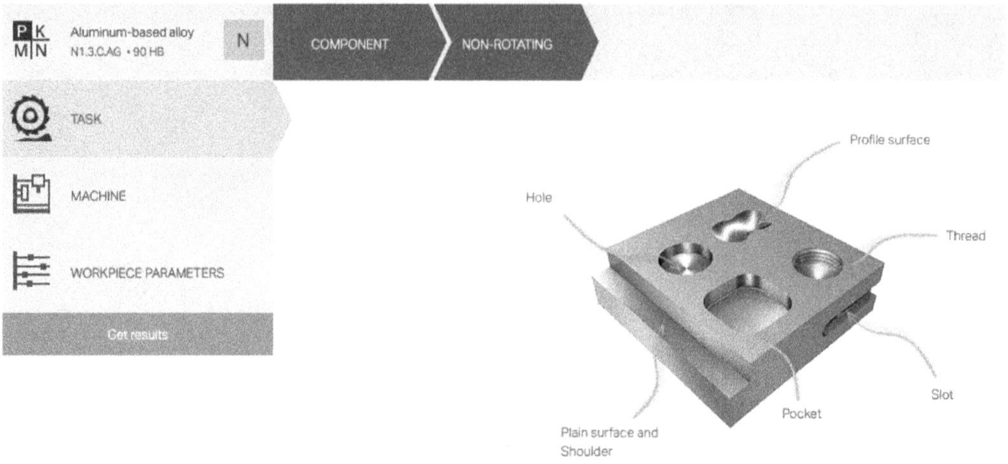

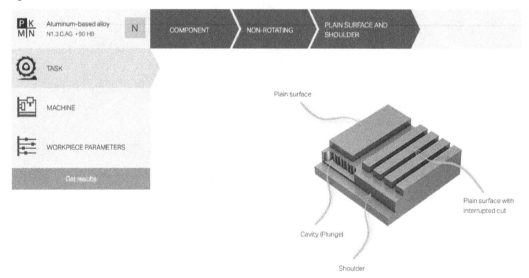

Figure 9.3: CoroPlus operation selection

There are a variety of categories we can select, but since we have to machine a flat surface with our facing operation, we shall pick **Plain surface and shoulder** (as you may recall, face milling and shoulder milling have many features in common, and therefore, they often are listed together).

Then, CoroPlus requires us to get even more specific with details. As you can see, we can find the **Plain surface** and **Plain surface with interrupted cut** options. These are the only two facing operations available.

Figure 9.4: CoroPlus operation selection

Plain surface with interrupted cut is better optimized for uneven surfaces with lots of interruptions on the path that may lead to pulsating loads on the tool. However, **Plain surface** is the one we shall pick right now; our stock is perfectly smooth at the moment, since it is still unmachined.

It is now time to specify the cutting operation we plan to implement:

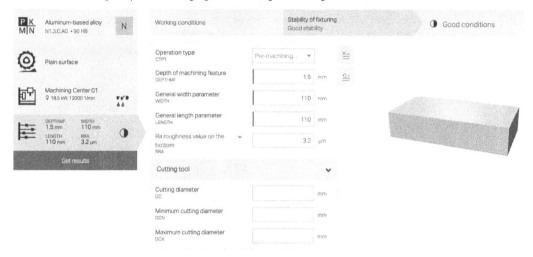

Figure 9.5: CoroPlus facing operation details

Let's go through the options that we need to consider:

- With **Operation type**, we can specify whether we want to create a roughing operation (called **Pre-machining**) or a roughing *and* finishing operation (called **Pre-machining and finishing**). Just for the sake of completeness, we will implement both operations; therefore, we shall pick **Pre-machining and finishing** from the drop-down menu.

- The layer we need to remove from the top of the stock is 1.5 millimeters, so we set that value for the **Depth of machining feature** option.

- Our stock is 110 x 110, so we can enter those values in **General width parameter** and **General length parameter** respectively.

- A useful parameter to set is **Ra roughness value on the bottom**; since our 2D drawing requires a roughness of 3.2 mm, we can specify that as well.

> **Note**
>
> Roughness is quite a complex attribute to understand and measure. To have an idea of what roughness is, we can say that it is a way of expressing a value related to the microscopic imperfections of a surface. As you may imagine, we cannot simply take a ruler and measure the roughness of a part. To do it properly, we need multiple measurements, along with a bit of math and statistics. There are several approaches to measuring roughness; here, we find **roughness Ra**, and in *Chapter 10*, we will mention **roughness Rz**. However, we don't have time to cover them in this book. If you want to read more about them, check out `https://en.wikipedia.org/wiki/Surface_roughness`.

- We do not have a **Maximum cutting diameter** value to specify, since we shouldn't have any leftover radius, as we are machining the entire face.

Now, we can simply hit **Get results** to let CoroPlus evaluate the best cutting strategy:

Figure 9.6: The first facing proposal

As we can immediately spot from the tool preview, CoroPlus is suggesting a typical facing tool with a KAPR of 45°, which is perfect for face milling; however, there is a catch.

As we can see from the picture, right after the face milling operation (**#1**), we have to create a circular pocket in the center of the stock (**#2**). Therefore, using the same tool for both operations may reduce machining time by avoiding tool changes.

Figure 9.7: The shoulder operation

A tool with a KAPR of 45° is perfect for face milling, but it is not capable of milling the central slot. Instead, we want to machine the highlighted area (**#2**) with the same tool; therefore, we shall reject any tool with a 45° KAPR and pick one with a 90° entering angle (KAPR).

Moving back to *Figure 9.6*, on the right, we can find a list of alternative tools we can use. These tools have not been suggested as the first choice, since they feature a less-than-ideal optimization for the operation we configured in CoroPlus.

We want to look for the first one with a 90° KAPR; it won't be the best one for face milling, but it will save us a tool change. Scrolling through the list of alternative tools, the most appropriate one is called **R390-040A32-11H**. You can find out more about it here: `https://www.sandvik.coromant.com/it-it/products/pages/productdetails.aspx?c=r390-040a32-11h`.

Since this tool has replaceable inserts, we also have to pick inserts as well. We have a large choice of inserts for this tool, but I would choose the following model, which is specific for aluminum alloys – **R390-11 T3 08E-NL H13A**. You can find more details on these inserts at the following link: `https://www.sandvik.coromant.com/it-it/products/pages/productdetails.aspx?c=R390-11%20T3%2008E-NL%20%20%20H13A`.

> **Note**
>
> Once upon a time, tools and inserts were squared and made of **High-Speed Steel** (HSS); today, there is always a wide range of geometries, materials, and coatings to select from instead. Every geometry, material, and coating may be better in certain situations with certain materials. In our case, for example, we picked the insert that offers the best performance when milling aluminum.

To re-evaluate the cutting parameters with this new tool and insert them, we simply click on the tool from the list:

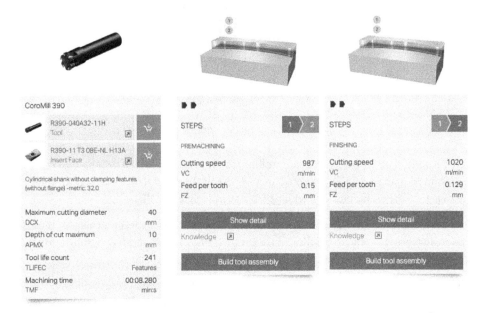

Figure 9.8: The new facing operation

As you can see, the same operation performed with this different tool (**R390-040A32-11H**) is taking around 8 seconds, instead of 6 seconds with the previous tool (**RA245-076R25-12H**). However, since a tool change would require way more time than 2 seconds, we should definitely opt for a common tool for both operations; this way, the productivity will increase, since the overall time will be reduced!

Now that we finally defined the tool to use, we can look for cutting parameters. If we click on **Show detail**, we can find even more pieces of information we will need to set later inside Fusion 360:

VC [m/min] CUTTING SPEED	FZ [mm] FEED PER TOOTH	N [1/min] SPINDLE SPEED	VFM [mm/min] FEED SPEED AT MACHINED DIAMETER
1 987	0.15	7850	7070
2 1020	0.129	8110	6300

AE [mm] WORKING ENGAGEMENT	AP [mm] DEPTH OF CUT	NOPAE NUMBER OF PASSES IN AE DIRECTION	NOPAP NUMBER OF PASSES IN AP DIRECTION
1 36.67	1	3	1
2 36.67	0.5	3	1

PPC [kW] CUTTING POWER	MMC [Nm] CUTTING TORQUE	HEX [mm] MAXIMUM CHIP THICKNESS	QQ [cm³/min] MATERIAL REMOVAL RATE
1 4	4.86	0.15	259
2 1.98	2.33	0.12	115

LEGEND

1 Premachining

2 Finishing

Figure 9.9: The cutting parameter extended details

Please note that, this time, we spent a great number of pages on how to use CoroPlus; since all future operations will be more or less the same, we will skip repeating the entire procedure. However, if you don't want to go for CoroPlus, you can still use the formulas explained in *Chapter 6*; this will be a bit more difficult and will require a bit more trial and error, but it is definitely possible!

Implementing face milling with Fusion 360

It is now time to jump back to Fusion 360. We can find the command for facing (called **Face**) under the **MILLING** tab inside the **2D** drop-down menu.

Figure 9.10: The Face command

As we can see, there are many other commands in the list; we won't cover them all, since some are very specific. However, as we progress through the book, we will try to use as many of them as possible.

For now, we will start with the **Face** command. Now, let's explore the different tabs.

The Tool tab

The first panel is called **Tool**. Here, we have to specify most of the cutting parameters and the tool to use (please note that we won't cover tool creation and tool import, since we should already be quite skilled at it).

Figure 9.11: FACE's Tool tab

In the **Feed & Speed** section, there are many values to specify; however, most of them are connected. Therefore, changing one will modify one or more of the others as well. Here, we just need to set **Spindle Speed** to **7850** rpm and **Cutting Feedrate** to **7070 mm/min**, as suggested by CoroPlus, and Fusion 360 will fill the other values for us.

Geometry tab

The **Geometry** tab is quite slim; we simply have to specify the outer contour for our facing operation.

Figure 9.12: FACE's Geometry tab

Using **Stock Selections**, pick the top face of the stock we need to machine, as shown in the following figure:

Figure 9.13: The contour selection

We can leave the tab called **Heights**, as the default values should already be good to go, and move directly to the tab called **Passes**, which is full of useful options we need to tweak.

The Passes tab

Inside this tab, we have to specify the remaining cutting parameters and several options that will shape the generated toolpath.

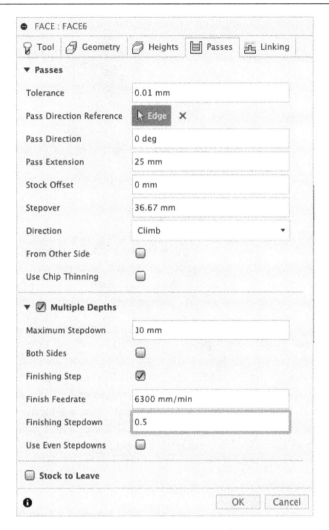

Figure 9.14: FACE's Passes tab

It is definitely worth a detailed look at the most important commands:

- **Pass Direction Reference** and **Pass Direction** allow us to set the direction of the table feed. Most of the time, these settings are not critical; however, we should always consider that at every cutting pass, the tool will extend beyond the stock contour. Therefore, we shall check that the tool won't collide with something else.

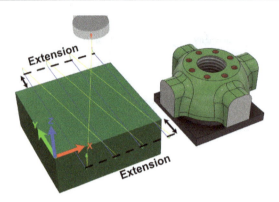

Figure 9.15: Pass direction

As you can see from the figure, on the right of the stock machined during the first placement, there is the stock that will be machined during the second placement. Therefore, choosing a cutting direction along the *Y* axis might be a good idea to solve the problem. For the **Pass Direction Reference** setting, we shall then pick any edge parallel to the *Y* axis and leave **Pass Direction** at 0°.

- **Pass Extension** and **Stock Offset** specify how much the tool will extend beyond the stock limit. Extending the cutting pass beyond the stock will help if we are unsure about the stock's dimensions, and we want to be sure to machine it entirely, even if it is a bit bigger than the ideal volume. In this case, I decided to add 25 millimeters for **Pass Extension** (but we probably could have used way less). Extending the toolpath via **Pass Extension** will increase the toolpath along the cutting direction only, while using **Stock Offset** can increase the toolpath in all directions. This time, I think we can safely leave the stock offset at zero.

- **Stepover** is a fancy way of referring to the radial cutting depth value; therefore, we can simply insert the calculated value of 36.67 mm.

- **Direction** allows us to specify the type of milling; we can opt for **Climb**, **Conventional**, or **Both**. Unless we are looking for maximum production speed, we should always opt for **Climb**, since it will give us less tool wear and better surface quality.

- **From Other Side** will flip the enter and exit points on our toolpath. There is no real need to use this option for our example; therefore, we will leave it untouched.

- **Use Chip Thinning** is used to create a curved entering path to the stock; it is useful to reduce tool loads and chattering. I don't plan to use it for the example part, but if you want to read more on the subject, I suggest you take a look at the following post: https://gorillamill.com/blog/chip-thinning/.

- **Multiple Depths** allows you to fine-tune the thickness of material removed by every cutting pass and introduce a finishing pass with a specific cutting depth. In our case, we will enable this option to get a better surface finish. Please note that we will use a separate finishing pass for training purposes only, since we may machine the upper face in only one pass.

- **Maximum Stepdown** is the maximum cutting depth permitted by our tool; in our case, it is 10 mm. This would have been a critical value to set if we had to machine a thicker layer of material.

- **Finishing Stepdown** is the cutting depth of the final finishing pass. In our case, we have to remove 1 mm of material as a roughing pass and a layer of 0.5 mm as a finishing pass; therefore, we only have to enter a value of 0.5 mm.

- **Finishing Feedrate** is the feed value for the final finishing pass only, which in our case is 6300mm/min.

There we have it – our first milling operation has been finally implemented. Now that we have machined the top face of our stock, we can focus on the remaining operations.

Shoulder milling

For our second milling operation – shoulder milling – we will machine the circular slot in the center of our part:

Figure 9.16: The slot profile

It has a diameter of 80 millimeters and a depth of 10 millimeters. As mentioned before, we will use the same tool already used for the first facing operation.

Calculating the cutting parameters by hand

Before launching the command on Fusion 360, we should evaluate the cutting parameters. However, the bad news is that this time, instead of using CoroPlus, we will try to calculate them on our own. Be sure of this – it will be a bit trickier but way more satisfying!

First of all, we should understand our machine specifications:

- We will use a Haas VF-1

- The maximum cutting power is around 22.4 kW

- Its maximum rated spindle speed is 8,100 RPM

- The maximum torque is obtained at 2,000 RPM with a value of 122 Nm

These values will be needed to check if our machine is powerful enough and can sustain the cutting parameters we calculated via the equations.

After that, we should briefly study the details of the cutting insert we are using. You can find all the necessary details using the following URL: https://www.sandvik.coromant.com/itit/ products/pages/productdetails.aspx?c=R390-11%20T3%2008E-NL%20%20%20 H13A.

From the URL, we have discovered that the insert is specifically designed for aluminum milling and that the suggested cutting values are the

$$f_z = 0.15mm$$

$$V_c = 990m/min$$

Here, f_z is the feed per tooth, and V_c is the cutting speed at the maximum diameter, which for our tool is 40 mm.

Please note that we should consider these values more like a starting point for our cutting operation. If for any reason our operation will lead to a bad result, we may need to tweak them a little bit.

Now, using the formula for cutting speed, we can get the spindle speed:

$$V_c = \frac{n\pi d}{1000}$$

We know V_c; therefore, we can flip the formula and get n:

$$n = \frac{V_c}{\pi d} * 1000 = 7878rpm$$

This value of the spindle speed is less than the maximum spindle speed granted by our machine (8,100 RPM); therefore, this is a good start. We can move forward to the next formula to calculate the table feed:

$$V_f = f_z * n * z_c$$

Where f_z is the feed per tooth, for which Sandvik suggested starting with a value of 0.15 mm, n is the spindle speed we just obtained, and z_c is the effective number of engaged teeth.

We already know that the actual number of teeth may change according to the geometry and radial cutting depth (if you don't remember why, please check *Figure 6.8*); however, to keep things simple, we will consider that all six teeth of our tool will be engaged all the time. This way, we will overestimate the required cutting power for a better safety margin. Therefore, if we insert the values in the previous equation, we obtain the following:

$$V_f = 0.15mm * 7878rpm * 6 = 7090.2 \ mm/min$$

Now, we have calculated V_f, and this is an important achievement, but we still need to solve one more equation, the cutting power equation. We can use the power equation to link all the cutting parameters together; this way, we can calculate unknown parameters.

Here, we can find the power formula in all its glory:

$$P_c = \frac{a_p * a_e * V_f * K_c}{60 * 10^6 * \eta}$$

These are the values we have to play with:

- a_p is the axial depth of the cut. The slot we have to machine is 10 millimeters deep, which is also the maximum axial cutting depth supported by our tool; therefore, we may decide to machine the entire slot with a single cutting pass!
- V_f is the table feed we just calculated.
- K_c is a value related to the specific cutting force needed to cut material at a certain feed. It is quite difficult to calculate properly, depending on the alloy, the treatment of the surface, and the cutting feed. However, for our example, we can consider it to be around 650 Mpa.
- η is a safety factor, used to consider friction and other variables that cannot be considered in formulas; most of the time, we can consider it to be between 0.7 and 0.8.
- P_c is the required cutting power for the operation with these parameters. It shouldn't be greater than our maximum spindle power (which, for the example, is 22.4 kW).
- a_e is the radial cutting depth (which we are currently missing).

Let's write down all the values we know inside a formula:

$$22.4kW = \frac{10mm * a_e * 7090\frac{mm}{min} * 650MPa}{60 * 10^6 * 0.75}$$

As you can see, there is only one missing parameter, a_e; therefore, we can calculate it.

If we do the math, we get a_e as 21.87 mm! However, there is a problem with this result. Sure, it is a solution to the cutting power formula; however, it breaks the golden rule of milling!

If you remember *Figure 6.13*, we should never engage the tool at a radial cutting depth close to half the diameter; a much better idea is to set it to 1/3 (13.33 mm) or 2/3 (26.67 mm).

Since a value of 26.67 mm is higher than 21.87 mm, we cannot use it, as our machine won't have enough power to cut with the other specified parameters.

Using the cutting power formula, we can reach the following conclusions:

- We can reduce the axial cutting depth to 8.2 mm and machine with a radial cutting depth of 26.67 mm

- Alternatively, we can use an axial cutting depth of 10 mm (the maximum for our tool) and a radial cutting depth of 13.33 mm (with an overall required power of 13.7 kW)

The first option may be better if we plan to implement a roughing operation that will remove 8.2 mm and a finishing operation that will remove 1.8 mm. However, if we check the part drawing presented in *Figure 8.1* in *Chapter 8*, we will find that there isn't a particular tight tolerance on dimensions or the surface finish on the sides or bottom of the circular slot we are about to machine. Therefore, I would opt for the second option – machining at a higher axial cutting depth; this way, we can save time machining in a single pass only.

> **Note:**
> Since this is a blind slot, we shall also calculate the cutting parameters used when the tool is plunging inside the stock. We will omit this part here, however we are going to mention helical movements in *Chapter 10*.

Now, we should also calculate the required cutting torque with the following formula:

$$M_c = \frac{P_c * 9549}{n} = \frac{13.7KW * 9549}{7878\ rpm} = 16.6Nm$$

To check whether this value is still plausible at 7,878 RPM, we would need a torque diagram (such as those found for vehicles); however, since this value is well below the maximum rated value, I think it is worth a first attempt at machining!

Please note that these formulas are considered a rule of thumb, and they may lead to a few trials and errors. If we get excessive tool bending or other issues while milling, we may have to adjust the parameters a bit.

That was the last formula we needed, so we can now move back to Fusion 360!

Shoulder milling inside Fusion 360

There are a couple of commands that can be used for shoulder milling inside Fusion 360; the first and most used is called **2D Adaptive Clearing**, while the other is called **2D Pocket**.

> **Note**
>
> If you find a command labeled **Adaptive**, it usually refers to the fact that such a command generates a better cutting path with an even tool load during the entire cutting operation.

You can find these commands in the following panel:

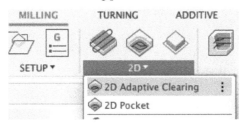

Figure 9.17: The 2D Adaptive Clearing command

Both of them are very similar; however, we will use **2D Adaptive Clearing**.

Tool tab

The first thing to do after launching the command is to fill in all the calculated cutting parameters in the **Tool** tab:

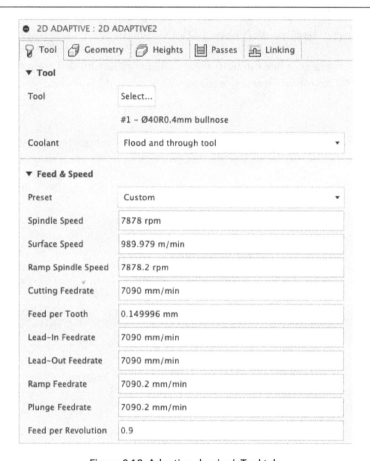

Figure 9.18: Adaptive clearing's Tool tab

After picking the same tool we used for the previous milling operation, we simply need to set **Spindle Speed** as **7878 rpm**, **Cutting Feedrate** as **7090 mm/min**, and **Feed per Revolution** as **0.9** mm.

After this, we can move directly to the next tab called **Geometry**.

The Geometry tab

In this tab, we have to specify the geometries to be included for toolpath generation, hence its name.

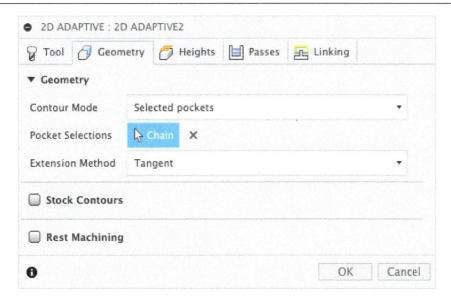

Figure 9.19: Adaptive clearing's Geometry tab

Using the options available in this panel, we have to specify the working area for our shoulder milling operation. With **Contour Mode** and **Pocket Selections**, we have to manually specify every face to be machined one by one. In our example, this shouldn't be a problem, since we only have one pocket to be machined; all we have to do is click on the contour of the cylindrical face, as shown in the following figure:

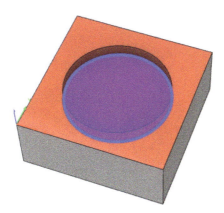

Figure 9.20: Pocket selection

Then, with **Extension Method**, we can specify how to approach open geometries and generate the toolpath around the faces. This is quite an advanced option and quite difficult to understand. However, since our pocket is closed, we don't really have to specify anything; we can leave the default setting to **Tangent**.

The **Stock Contours** option allows us to specify a restricted working area to avoid our tool moving into unwanted areas. Since the pocket we are about to machine is closed and completely inside the stock bounding box, we can leave this option unchecked.

Rest Machining is a useful option that permits us to machine areas where a previous operation couldn't remove material. For example, let's imagine we implemented a roughing pass using a tool with a big diameter. The finishing pass will be performed with a much smaller tool. Enabling this option will allow the tool to machine areas untouched by previous operations. Once again, this isn't applicable in our case; therefore, we shouldn't set this flag either.

We can leave the **Heights** panel with the default values and move on to the **Passes** panel for the last parameters to be set.

The Passes tab

Inside this panel, we can finalize the cutting passes for our operation:

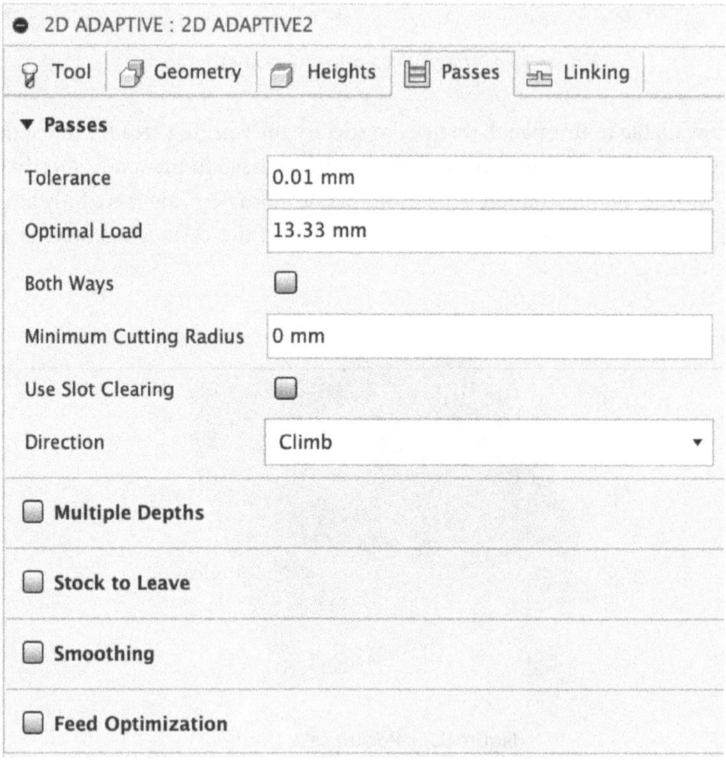

Figure 9.21: Adaptive clearing's Passes panel

We will now review the most important options available here:

- **Optimal Load** is the value of our radial cutting depth, which is 13.33 millimeters. Please note that this value is more of a suggestion that we are giving to Fusion 360. It will try to respect this value; however, due to geometry or tool shape, it may be forced to use a different value when machining certain areas!

- **Both Ways** is an option that allows us to specify whether we intend to use both climb and conventional milling, however you should never go for this option; please remember, as discovered in *Chapter 6*, to always use climb milling if possible.

- **Minimum Cutting Radius** is an advanced option to convert every sharp corner inside the tool path into a softer curve with a fillet radius. Whether our technology is milling, laser cutting, or 3D printing, sudden changes in direction along the toolpath (for example, on corners) are always to be avoided if possible, since they can cause vibrations. With this option, we can automatically convert sharp corners into tiny smooth fillets. For our example, there aren't sharp corners; therefore, we don't have to tick this flag.

- **Use Slot Clearing** is an option that we should always check when machining a small pocket (or slot) with tight tolerances. We already covered the idea behind this type of operation in *Chapter 6*, which you can see in *Figure 6.15*.

- With **Direction**, we can specify whether we intend to use **Climb** milling or **Conventional** milling; we should always leave the default value set to **Climb**.

- **Multiple Depths** allows us to specify several cutting passes and an optional finishing pass. Since we are planning to machine the entire slot at an axial cutting depth of 10 mm, we will machine it with a single pass.

- **Stock to Leave** is the option that allows us to leave a thin layer of material above our geometry for future machining; in our case, we don't need it.

Moving on from the **Passes** tab, the **Linking** tab contains many options related to toolpath optimization when moving from one cutting pass to the other; however, it is a bit too advanced and we won't cover it. Don't worry – most of the time, we can leave the default settings untouched! We will approach this advanced tab in *Chapter 10*.

Now that we have filled all the required parameters and options, we can check the resulting simulation; at this point, we have the first two milling operations completed:

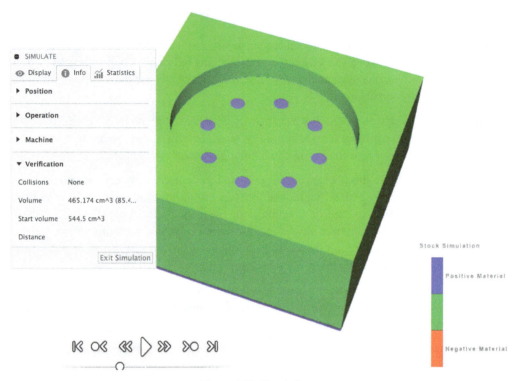

Figure 9.22: Simulation

As we can find from the preceding figure, the machined area is green, meaning that we removed the right amount of material from the stock. From the **Verification** tab, we can see that no collisions were found. Everything seemed to go as intended. We now just need to drill the eight holes and machine the threads, which we will do in the next section.

Drilling

We fully covered the drilling operation in *Chapter 5*, for our turning example. The command is the same for milling as well; therefore, we will skip a long introduction regarding the panels and just cover the required settings for the example.

So, as before, let's look at the **Tool** tab first.

The Tool tab

Inside this panel, we can specify the tool to use and the main cutting parameters:

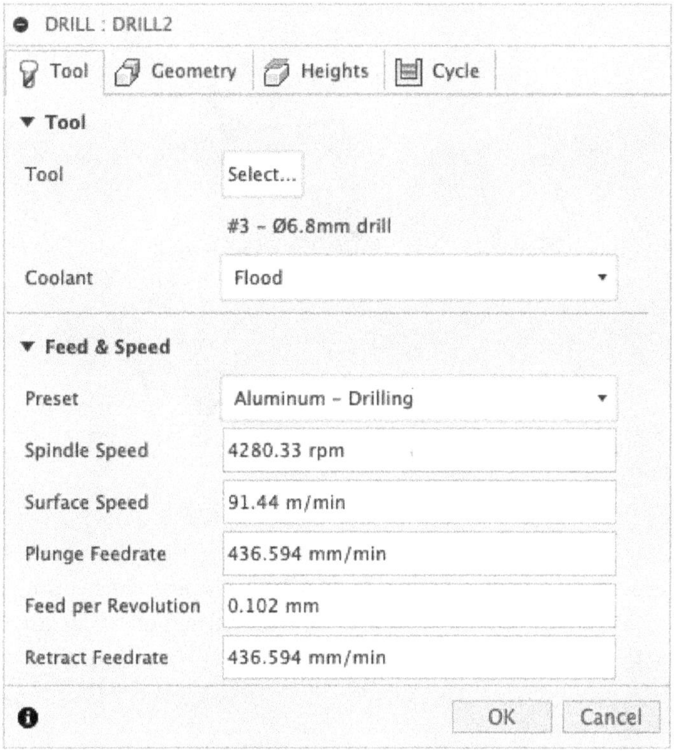

Figure 9.23: Drilling's Tool tab

The first thing we have to set is the cutting tool, which for a drilling operation is a drill bit. Since the holes will need to be threaded, we have to create a pilot with a certain diameter. The thread we will have to tap is an M8; therefore, the pilot hole shall have a diameter of 6.8 mm.

So, as you can see from the preceding screenshot, we have to pick a drill bit with a diameter of 6.8 mm.

Now that we have chosen the right tool, we can focus on the cutting parameters. Clicking on the drop-down menu called **Preset**, we can pick a suitable set of parameters for drilling operations according to the material; in this case, we can select **Aluminum - Drilling**.

The Geometry tab

Inside this tab, we have to pick the geometry to be machined; therefore, we shall select every hole on the 3D model.

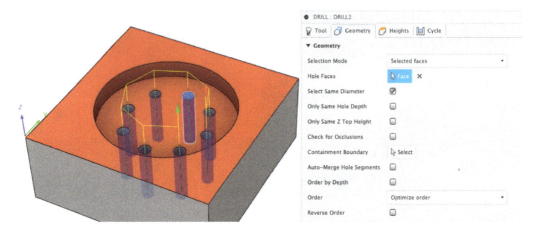

Figure 9.24: Drilling's Geometry tab

If we don't want to select every hole one by one, we can select the first one by using **Hole Faces** and then flag the option called **Select Same Diameter**. Fusion should automatically discover all the holes to machine; however, this option may fail if the geometry is complex or non-cylindrical.

The Heights tab

In this tab, we can find many options to set the working height of the tool when it is outside the part (for example, the clearance height or the retract height), but we can also specify how deep the drill bit will go when plunging inside the stock.

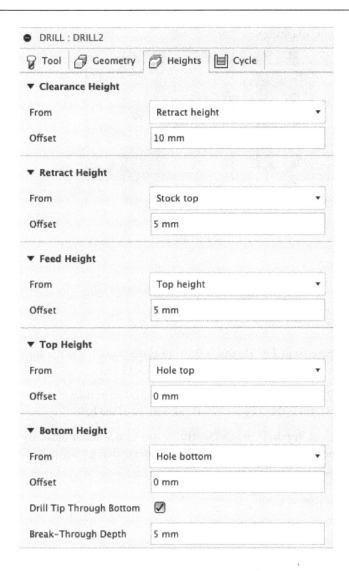

Figure 9.25: Drilling's Heights tab

As you can see, there is an option called **Drill Tip Through Bottom**. What is that all about? The origin of drilling tools is on the tip of the tool; therefore, it will move to the bottom of the hole, resulting in a bad hole shape. If we want to create a hole that goes straight through the part, we have to make sure that the tip of the bit will move below the bottom of the hole.

Let's look at the following diagram:

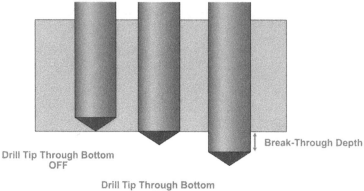

Figure 9.26: Drilling options

In this diagram, we can see the following:

- On the left, where **Drill Tip Through Bottom** is disabled, the drill bit just came up to the bottom of the hole; once retracted, the result is a blind hole.

- In the center, **Drill Tip Through Bottom** is enabled, and the hole should result in a perfectly cylindrical shape, since the tip goes below the bottom of the hole.

- On the right, using **Break-Through Depth**, we specified a minor offset for extra safety to be sure the tool goes through the entire stock. This is the option we will pick, along with an offset of 5 millimeters.

It is now time to take a look at **Cycle**.

The Cycle tab

Inside this tab, we can specify the type of drilling operation we want to implement and its settings.

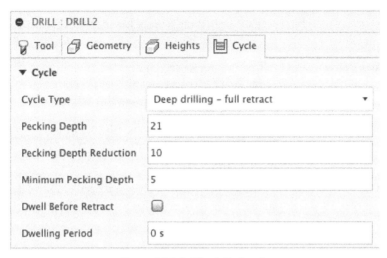

Figure 9.27: Drilling's Cycle tab

Since we have to drill the pilot holes, as **Cycle Type**, we could go for **Drilling - rapid out** (the same operation we implemented in the turning example). However, since the holes are quite small and deep, I'd rather pick **Deep drilling - full retract**. With this operation, we have much more control over how the drill bit will plunge into the hole; in particular, we can specify pecking settings.

Pecking is a chip management strategy particularly useful for deep and small holes. That's because, with this type of drilling, there is a high amount of chip produced.

When the hole is very narrow and deep, it gets more difficult to remove the chip. The reason is quite simple – the chip will be more prone to getting trapped at the bottom of the hole.

With pecking, we can drill only a fraction of the hole and then retract the tool to evacuate most of the chip stuck inside the hole. After the first retraction, the drill bit will plunge again into the hole and keep drilling, without the bother of excessive amounts of chip.

Looking back at *Figure 9.27*, **Pecking Depth** can be adjusted according to the tool and the depth of cut. As a rule of thumb, we can drill a hole up to a depth of around three or four times the drill bit diameter without any issues. Therefore, in our example, I chose a **Pecking Depth** setting of 21 mm.

Then, the more the hole depth increases, the more we need pecking, and the more we have to reduce the pecking distance. To control pecking passes according to the drilling depth, we can introduce a **Pecking Depth Reduction** setting of around twice the diameter (a value of 10–14 millimeters may be a good first approach).

After a while, as long as the hole depth increases, we can no longer reduce the pecking depth, as the hole is simply too deep; therefore, we reach a **Minimum Pecking Depth** value that has to be maintained down to the bottom of the hole.

A typical value for such a minimum pecking depth is the same or less than the drill bit diameter. In our case, I think we should set a smaller minimum pecking depth, so I will go for a value of 5 millimeters.

With **Dwell Before Retract** and **Dwelling Period**, we can let the tool spinning at the bottom of the hole for a few instants; this is usually done for a better surface finish on the bottom of the hole. In our case, since we are machining through holes, we can leave these options unchecked.

There are many concepts to properly master pecking and dwelling, which we won't cover in this book; however, there is a very interesting video by Haas Automation on the subject that you should definitely take a look at: `https://www.youtube.com/watch?v=AM6nVgKjBQo`.

Congratulations! The first placement is almost completed, but we still need to thread the holes using a tap!

Tapping

There is not a dedicated command for tapping, so we have to use the drilling command instead. Since we just reviewed the command for drilling, we will just cover the important settings for tapping.

We have to thread the first 20 millimeters of the holes with an ISO M8 x 1.25 thread. To do this, we have to choose a tap with the same pitch and diameter. Luckily, there are many tapping tools at our disposal in the default tool library; therefore, we don't have to create or import a new tool.

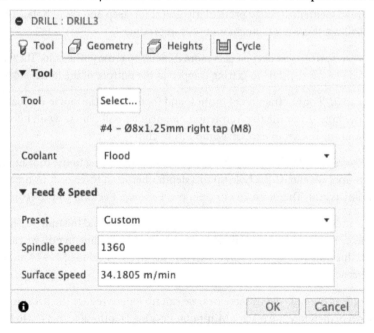

Figure 9.28: Drilling's Tool tab (used for Tapping)

CoroPlus suggests a **Spindle Speed** setting of `1360` RPM, resulting in a **Surface Speed** setting of `34.1805 m/min`, so all we have to do is input those values inside the panel.

Then, we have to specify all of the holes to machine using the **Geometry** panel; we won't repeat this procedure, since it's identical to how we drilled the pilot holes in the last section.

However, we should go to the **Heights** panel, where we can specify the thread length.

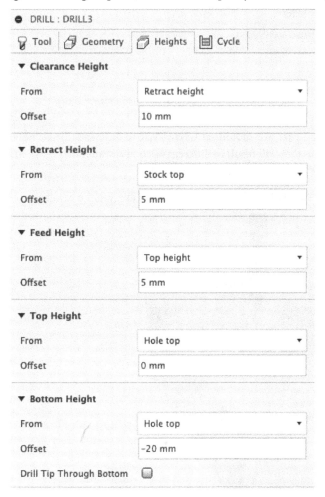

Figure 9.29: Drilling's Heights tab (used for Tapping)

The planes specified inside this panel define the working boundaries for the cutting operation; therefore, in order to set the maximum tapping depth (thread length), we have to set the **Bottom Height** plane at the maximum level to be reached by our tap. In particular, we need to specify the bottom plane to be measured from the hole top and set an offset of **-20 mm**.

After setting the working boundaries, we can move to the **Cycle** tab:

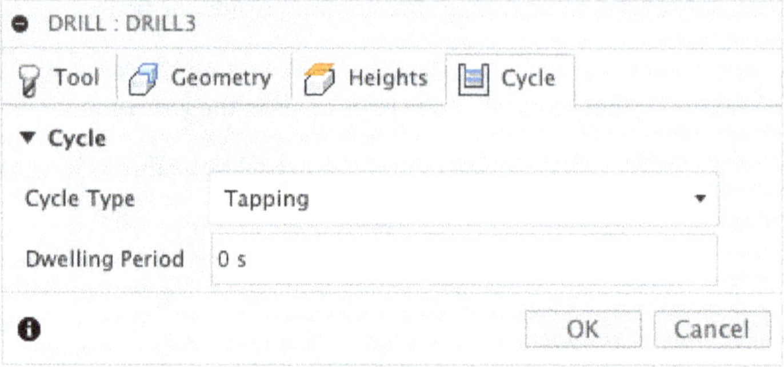

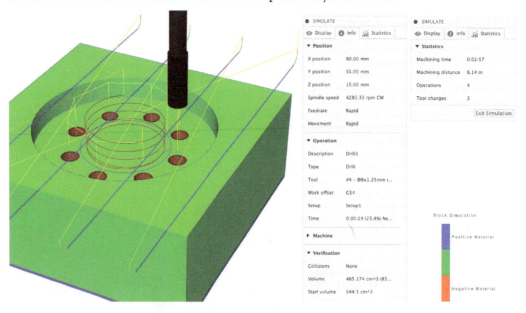

Figure 9.30: Drilling's Cycle tab (used for Tapping)

Since we are threading a part, we should set **Cycle Type** to **Tapping**. Please note that there are also other tapping operations that will result in different G-code commands; you should pick the one that fits your CNC machine and its post-processor the best.

It is now time to run a simulation of the entire setup and analyze the results:

Figure 9.31: Simulation

First of all, inside the **Verification** panel, we can find that no collisions occurred, which is always a good start. Then, we can find that every face of the stock is painted green, so we know that we machined the right amount of material from the stock.

As you may have noticed, the holes are depicted in red. Is it an error? Don't worry – it isn't. When modeling using Fusion 360, holes have the dimension of the pilot hole (in our case, 6.8 millimeters).

Once the tap enters the hole to machine it, it will increase its diameter up to M8; therefore, the simulation results in a warning of negative material on the face and it will display it in red. To cut a long story short, Fusion 360 thinks that we removed too much material, since the original pilot hole was 6.8 mm, but in reality, we simply threaded the holes to M8.

Another critical point to check is the **Work offset** setting, which at the moment is set to G54, which is a reference to the origin of the first placement (as discussed in *Chapter 8*). When simulating the next placement, we should check the that work offset matches the origin of the second placement.

Also, as an interesting side note, we can check from the simulation panel that the entire machining took a little more than 2 minutes and we had to use three tools. This is useful to calculate the costs for the part in relation to machining time.

Everything is looking fine; therefore, we can move confidently to the next stock placement, where we will machine more complex geometries.

Summary

That concludes this chapter. I hope that the extra effort required for the equations was worth it; now, we don't really need online tools anymore (even if they are still valuable).

In summary, we first set up face milling using CoroPlus. Then, we decided to change the suggested cutting tool and opt for one capable of machining the round pocket in the center of the stock.

After the first facing operation was implemented in Fusion 360, we moved to the circular slot in the center of the part and calculated all the required cutting parameters, using the milling equations.

After completing the milling operations, we moved to drilling and tapping. First, we found out how to drill holes using the same drilling command already introduced in *Chapter 5*, but this time, we mentioned a chip evacuation technique called pecking. After drilling the holes, we moved on to tapping and found out how to set the thread length and check for errors in the final simulation.

In this chapter, we didn't implement fancy 3D contouring operations. Instead, we started with basic operations such as face milling and shoulder milling; however, these milling strategies are really common and really powerful.

Now, follow me to the next chapter, where we are about to complete the example part by machining the second placement of our stock!

10

Machining the Second Placement

This chapter directly follows on from the previous one since we will start right where we left off before. Now that we have completed all the milling operations on the first placement (the first setup), we can focus on the second placement by implementing all the remaining operations that will result in the final shape of the part.

What we are about to do is clearly depicted in the following figure:

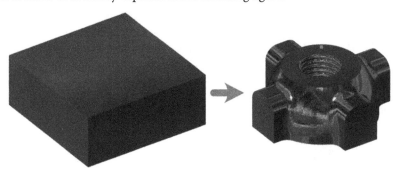

Figure 10.1: The second placement

The partially machined stock has now been flipped and installed onto the fixture we mentioned in *Chapter 8*. With this second placement, we will machine the opposite side of the part. We will now cover every remaining cutting operation one by one.

The goal of this chapter is to introduce more complex operations than those found in *Chapter 9*, leading you closer and closer to a real case scenario with an ever-increasing level of complexity. We will also try to analyze minor details that may hide potential issues if not handled correctly.

In this chapter, we will cover the following topics:

- Face milling
- Implementing a roughing operation using adaptive clearing
- Milling a hole
- Finishing the part using a morphed spiral
- Thread milling

Technical requirements

In order to follow this chapter and get the most out of it, you should have read all the previous chapters and have a clear idea of the milling operations introduced in *Chapter 6* and *Chapter 9*.

Since we already covered several ways of calculating the cutting parameters, we won't focus on how to get them anymore; therefore, you should already be capable of calculating the parameters and choosing the right tool yourself. Of course, it is also possible to simply copy and paste the tools and the values used here.

Face milling

The first milling operation we should set up is, once again, face milling; in this case, we have to remove a layer of 1.5 mm on the top of the stock. We won't cover this operation here since it is identical to the first one we introduced in *Chapter 9*; therefore, we may keep using the same tool and the same cutting parameters.

After the first facing, we should find something similar to this:

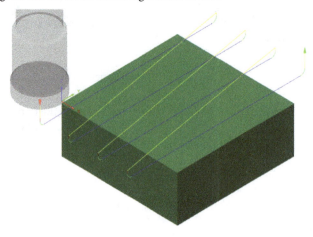

Figure 10.2: Face milling

As we can see, the diagram displays both the cutting tool and the generated toolpath for the first face milling I implemented. We already know every setting from *Chapter 9*; therefore, I won't bother you with a recap of what we did last time. However, this time I changed something in the settings. Let's try to discover what I changed just by looking at the toolpath:

- First of all, we can find an entry point close to the origin of the coordinate system, highlighted by the red cone pointed downward (to the left of the image), and an exit point highlighted by a green cone pointed upward (to the right of the image).

- Then we can spot that the cutting operation is performed with four parallel movements performed in the same direction; from that, we can conclude that the milling tool will most likely machine the stock with up milling only or down milling only and not both (if you need a recap on the differences between these types of machining, please go back to *Figure 6.7*). We can affirm this since the tool, after completing every cutting pass, retracts to a safe height and repositions to perform the next cutting pass in the same direction.

- In addition, we can identify that the cutting movements extend quite a bit beyond the stock limit. Up to this point, we cannot spot any difference from the face milling operation in *Chapter 9*.

- The only difference we can spot from the toolpath is the number of cutting passes. Since there is only a single set of blue lines, we can deduce that there is not a roughing pass and a finishing pass, but the entire layer of material removed is machined in just one pass.

That was not bad for just a glance at a preview picture!

We can now move on to the next milling operation needed to roughly machine the 3D shape of the part for future finishing operations.

Implementing a roughing operation using adaptive clearing

Now that we have machined the top of the stock, it is time to move on to a roughing operation on the complex shape of our part. The goal of this operation is to remove most of the material around it and leave just a tiny bit for the next finishing operation.

Long story short, we have to create something similar to a South American pyramid:

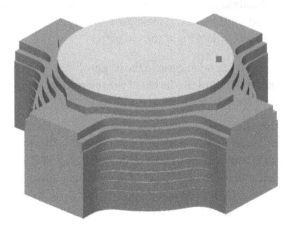

Figure 10.3: Steps on the side of our part

As you can see, there are multiple steps all along the sides of the part. The bigger the steps, the faster this roughing operation will be. However, please remember that a finishing operation is not capable of removing much material; therefore, we should find some sort of sweet spot between optimizing the roughing passes and the finishing passes. For this operation, we will aim for each step to be 2 mm.

Before starting with the command and settings, we should ask ourselves what the best tool to pick is and whether there are certain constraints it should respect. From my point of view, there aren't many hidden troubles in the geometry, so we should simply pick a tool long enough to machine up to the bottom of the part (which is 42 mm tall).

Deciding the tool diameter is a bit trickier; the diameter should be the biggest possible yet small enough to fit the internal fillets of the part. In the following diagram, we can find the fillets we have to machine with our tool:

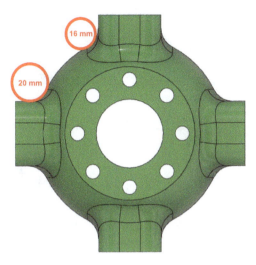

Figure 10.4: Fillets

As you can see in the diagram, a tool with a diameter of 20 mm would barely fit the shape of the part; therefore, since we also have to leave a bit of material for the finishing operation, we should pick a smaller tool. In this case, I will go for a 16 mm tool: it perfectly fits the part, and it is still quite a strong tool capable of heavy loads and a high material removal ratio. Of course, we may also pick smaller tools, but they will probably be less effective.

The tool I decided to use for this operation is called **1P260-1600-XA 1620**, and it is a carbide end mill particularly suitable for heavy roughing. You can find more details at the following link: `https://www.sandvik.coromant.com/en-gb/products/pages/productdetails.aspx?c=1p260-1600-xa%201620`.

Why did I choose this tool? There are multiple reasons; first of all, its cutting diameter is 16 mm, it is a heavy-duty tool capable of a high material removal ratio, which is a perfect fit for a roughing strategy via shoulder milling, and finally, it has a maximum axial cutting depth of 50 mm, meaning that we may machine the entire height of the stock in one single pass.

> **Note**
>
> Though this tool is not specific for aluminum alloys, it should work just fine. However, there may be more optimized solutions in the catalogue!

Now that we have a general idea of what type of operation and tool we are about to use, we can finally launch the command that will allow us to implement this roughing operation. Since we are about to machine a complex 3D surface, we will need a 3D milling strategy. 3D milling operations can be found in the following panel:

Figure 10.5: 3D milling commands

Don't get distracted by the large number of operations available in the list; the only two roughing commands are **Adaptive Clearing** and **Pocket Clearing**. They are similar and can be used for the same type of geometry. The main difference between the two is hinted at by the term "adaptive." An adaptive command is always preferred over standard cutting strategies since it is capable of generating more advanced tool paths for a constant load on the tool.

In our case, we will click on **Adaptive Clearing**, and then we'll go through each of the panel's tabs.

> **Note**
>
> I won't bother you with cutting parameters calculation anymore since we should now have the resources to calculate all of them by hand or using online calculators. I will only recap the most important ones so you can use the same parameters I picked (it is up to you to guess whether I used CoroPlus or if I did it all myself).

The Tool tab

After clicking **Adaptive Clearing**, the first tab you will see is the **Tool** tab:

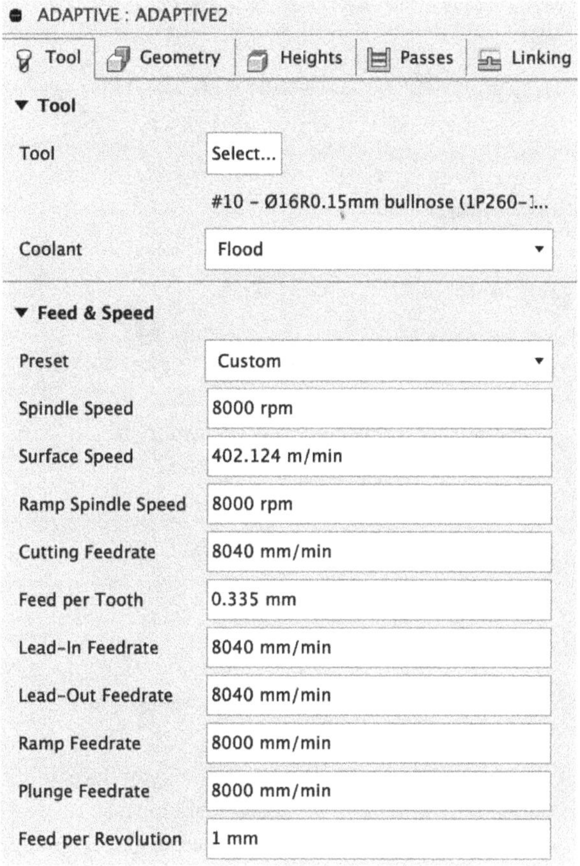

Figure 10.6: The Adaptive Clearing Tool tab

These are the settings for our starting point:

- The **Spindle Speed** we are going to use is **8000 rpm**; therefore, the cutting **Surface Speed** is **402 m/min**

- **Feed per Tooth** is **0.335 mm**, resulting in a **Cutting Feedrate** of **8040 mm/min**

The Geometry tab

There is not much we have to modify in the **Geometry** tab; the only option we need to look at is **Rest Machining**.

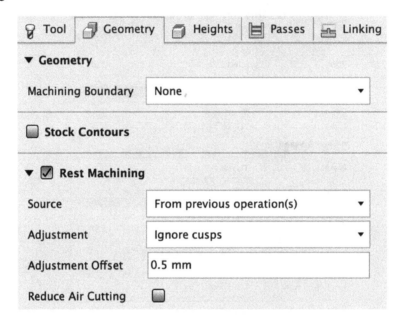

Figure 10.7: The Adaptive Clearing Geometry tab

With **Rest Machining** checked, our tool won't try to machine the top face of our stock again (in fact, we already removed the material there with our first facing operation). Not only will this reduce wasted time, but the tool won't risk rubbing against an already machined surface, shortening the tool's life span!

The Heights tab

For the **Heights** tab, you can leave the fields untouched and just use the default values, as you can see in the following screenshot:

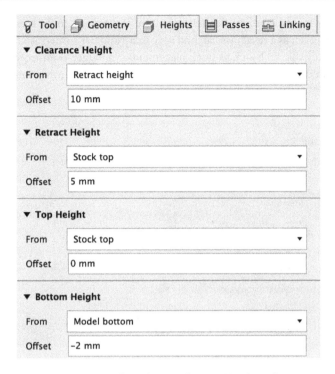

Figure 10.8: The Adaptive Clearing Heights tab

However, I would suggest tweaking the **Bottom Height** plane a little bit.

The **Bottom Height** plane is our operation's lower boundary on the z coordinate. Adding a couple of millimeters below the bottom plane is a safe way to ensure that all the material is removed.

For example, let's suppose that the stock is slightly thicker than the theoretical value. In this scenario, the machine has no way of figuring it out; therefore, it will machine the sides of the stock up to the theoretical bottom of the part. In reality, the true bottom of the part is a bit below the theoretical bottom level; if this happens, there is the risk of finding sharp faces on the bottom of the part.

The following figure demonstrates the problem:

Figure 10.9: Bad results

As you can see, our tool moved up to the theoretical bottom of the stock, in the exact position where it should have ended; however, as a result, we can find very sharp and flat corners left from the operation. Those are most probably caused by a bad stock placement or a bigger-than-expected stock height. Therefore, imposing an offset to the bottom of the part is a good idea to avoid this type of issue.

The Passes tab

Next up is the **Passes** tab:

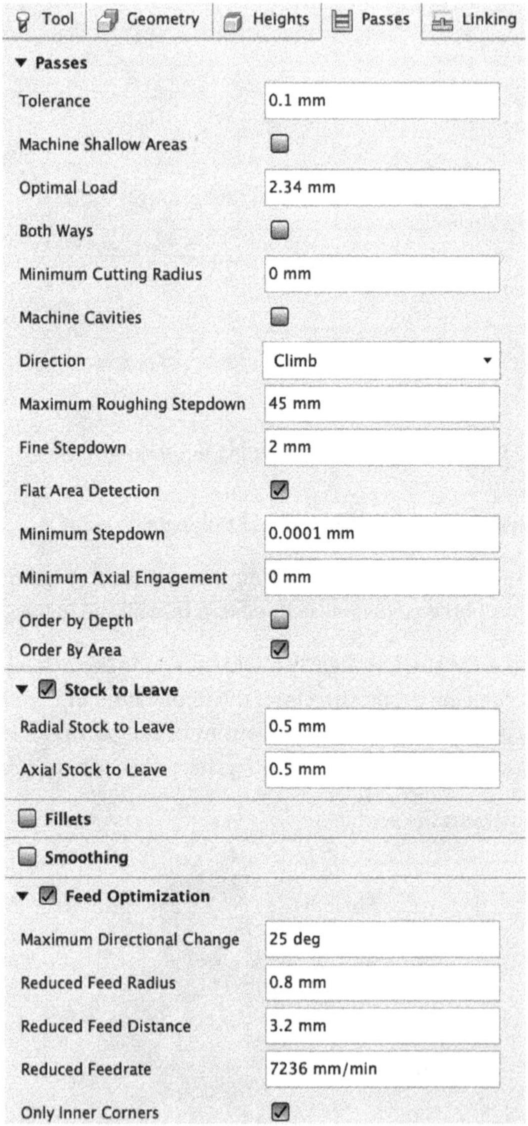

Figure 10.10: The Adaptive Clearing Passes tab

Once again, this tab is one of the most content-dense and includes some very important settings; these include the following:

- **Maximum Roughing Stepdown**: This parameter is the maximum axial depth of cut allowed. As mentioned, I plan to cut the sides of the stock with a single deep shoulder milling operation. Our theoretical stock height is 42 mm, and to this value, we added 2 mm as the bottom offset, resulting in a total required height of 44 mm. Our tool can cut up to 50 mm; therefore, any value bigger than 44 and smaller than 50 will do the job. In the example, I set a maximum stepdown of **45 mm**.

- **Fine Stepdown**: Similar to the previous parameter, this value sets the actual height of the steps of our pyramid. The value specified here is responsible for how rough the roughing operation will look once finished. Smaller values will lead to a higher machining time, but a faster finishing strategy and bigger values will result in a shorter roughing operation but in a much more demanding finishing operation. We should aim for a sweet spot, so I would try somewhere between 2 and 4 mm before settling on **2 mm**.

- **Optimal Load**: This is the radial depth of cut of every cutting pass. Since we plan to machine the sides of our stock at high axial depths, we have to set a tiny radial depth of cut. The value I'm going to use is **2.34 mm**.

- **Direction**: With this parameter, we can set the type of milling we intend to use; for our example, we are setting up quite a demanding cutting operation; therefore, I would definitely pick **Climb** milling for better tooth performance.

- **Both Ways**: Checking this option will result in a toolpath with climb milling areas and conventional milling areas. It is mainly intended for time-saving purposes. For our example, I will not check this option.

- **Stock to Leave**: Here, we can specify an extra layer of material all around our part. This extra material will be removed by the following finishing operation. For example, we may leave a layer around 0.5 to 1 millimeter thick.

- **Order by Depth** and **Order by Area**: These two options optimize the generated toolpath reducing retracting movements. For our example, we can say that there are four distinct areas at the four corners of our stock. Therefore, using **Order by Area**, we can impose Fusion 360 to completely machine each corner before moving to the next one. This type of behavior will result in a faster machining time. **Order by Depth** is the default behavior, and it will machine layer by layer, increasing the cutting depth at each cutting pass.

- **Feed Optimization**: Checking this option will reduce the table feed when approaching a sharp corner in the toolpath; this is useful to reduce stress on the tool and avoid chattering. Our shape is not really sharp; however, enabling this option is always a good idea, and it doesn't really affect machining time. We can leave the default values (between **Maximum Directional Change** and **Only Inner Corners**) unchanged.

- **Minimum Cutting Radius**: This is used to remove sharp corners on the toolpath; in our example, we won't need this as we don't have any corners to handle.

- **Machine Cavities:** With this option unchecked, we can let Fusion 360 generate a cutting toolpath on just the outer faces of the part and will machine the hole in the center as well. Since I plan to create a dedicated operation for the hole, we can leave this unticked. It is mostly used to implement different operations for outer contours and closed slots.

The Linking tab

We haven't explored the **Linking** tab in any of the previous operations we have implemented, but this time I just want to give you a glimpse of its usage. Inside this tab, we can find settings that are not really related to cutting movements, but rather, they manage the way the cutting passes are linked together (hence its name).

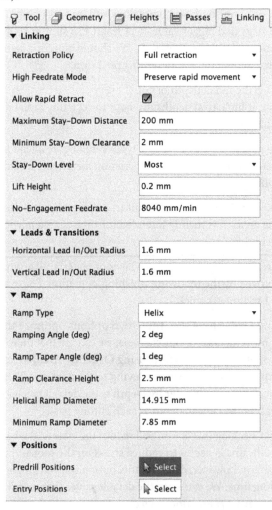

Figure 10.11: The Adaptive Clearing Linking tab

Most of the time, we can leave the default values; however, here, we are going to change the **Stay-Down Level** policy.

What is the **Stay-Down Level** policy? During the cutting process, sometimes the tool moves without cutting the stock. There are many situations where this may occur, for example, when the tool is repositioning between cutting passes. When not cutting, the safest approach is to completely retract the tool upwards; this way, we can be sure that no collisions occur during the rapid movements.

Retracting the tool, however, is not instantaneous; therefore, if there are many retractions, we may have long periods of unproductive time during the overall machining time of the part. In order to reduce the time wasted on retractions we can force the tool to stay at cutting depth as long as possible and reduce the retraction distance.

This behavior is controlled with **Stay-Down Level**. If we increase the value up to **Most**, we get the fastest machining time (but it is more dangerous for the tool due to unforeseen collisions); if we leave the default value set to **Least**, we waste a lot of time on **rapid movements**.

> **Note**
> Rapid movements are a particular G-code setting that forces the CNC to move in the fastest way possible when not cutting. Every movement where the tool is not engaging the stock should be set to rapid to reduce unproductive times.

In the following diagram, we can find the different toolpath generated with these two values (**Stay-Down Least** and **Stay-Down Most**):

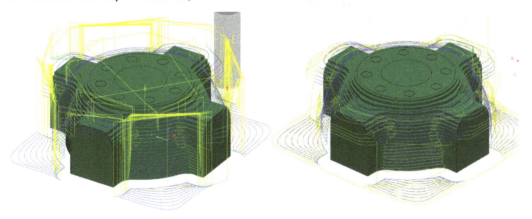

Figure 10.12: Stay-Down Least versus Stay-Down Most

As you can see, the toolpath on the left (generated with **Stay-Down Level** set to **Least**) has so many retract movements (depicted with the yellow lines), while the toolpath on the right with **Stay-Down Level** set to **Most** has much less retraction distance.

Back to our example, since the outer geometry is open and since the stock is held in place by a custom fixture that keeps the part above the working bed, I don't foresee any issues with chip evacuation or fixture; therefore, I'm quite confident in letting the tool stay at high depth as much as possible. Therefore, I'm going to set **Stay-Down Level** to **Most**.

Changing the **Stay-Down** policy greatly affects the machining time; there is a difference of 20% between the two examples, which for mass production is a huge value!

Also, please note that the tool is not going to rub against machined faces either since it will still retract by the value specified by **Lift Height**, in our case 0.2 mm.

Now that we have set and reviewed every parameter for the **Adaptive Clearing** command, we can now move on to the next operation where we will machine the big hole at the center of the part.

Milling a hole

It is now time to implement a machining strategy for the big hole at the center of our part. It's quite a big feature since it is a pilot hole for an M36x4 thread with a diameter of 32 mm. With such a diameter, it is impossible to use a drill bit; therefore, we have to create a milling operation instead.

In order to reduce tool changes, we are going to use the same tool already used for the previous operation. This type of tool is suitable for many different milling operations. There are multiple commands inside Fusion 360 to mill a hole; for example, we may go for another **Adaptive Clearing** operation.

However, I want to introduce you to as many commands as possible; therefore, we are going to use a command that we can find in the **2D Operations** set called **Bore**. This command is specifically used for machining cylindrical or conical features and is perfect for our case:

Figure 10.13: Bore

Once launched, you'll see a familiar panel. Like usual, let's take this one tab at a time.

The Tool tab

In the **Tool** tab, we have to insert most of the cutting parameters for the operation:

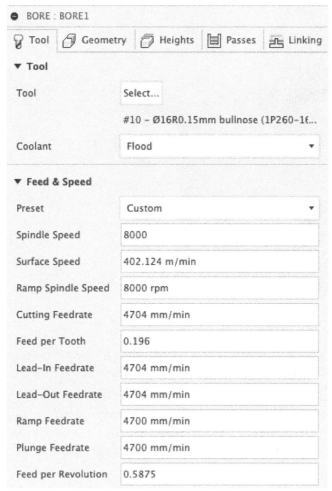

Figure 10.14: The Bore Tool tab

Here are the values that we will use:

- **Spindle Speed** is set at **8000**, which will lead to **Surface Speed** being **402.124 m/min**.

- Using the **Feed per Tooth** suggested for the tool of **0.196**, we get a **Cutting Feedrate** of **4704 mm/min**.

- The last value we can insert is **Feed per Revolution**, which can be calculated by dividing **Cutting Feedrate** by **Spindle Speed**. In our case, the result is **0.5875**.

Those are good values to use for the first attempt. However, since the hole we are about to machine is quite narrow and deep, we may encounter chip evacuation problems; if this is the case, we may need to use more conservative values.

The Geometry tab

Next up is the **Geometry** tab:

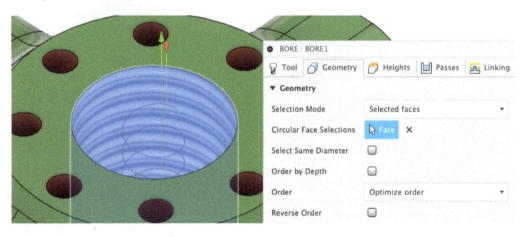

Figure 10.15: The Bore Geometry tab

There is not much we have to modify here; we simply have to select the cylinder surface using **Circular Face Selections**. This option allows us to pick any cylindrical face inside the part to be used for toolpath generation.

As already encountered in the drilling operation, we may let Fusion 360 automatically pick all the features with the same diameter by clicking **Select Same Diameter** and order the machining operations by depths using **Order by Depth**. As you have already guessed, those options are useless in the example since we have just one single hole to machine.

The Heights panel

Moving to the **Heights** panel, we can leave most of the parameters the same, apart from one parameter that we should already be familiar with: **Bottom Offset**, which we already mentioned for **Adaptive Clearing** and for **Drilling**.

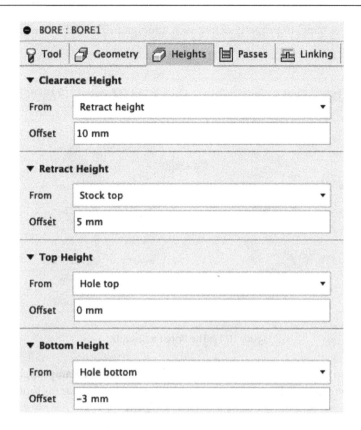

Figure 10.16: The Bore Heights tab

Since we want to be sure to machine the entire height of the hole, it is a good idea to add a small **Offset** from **Bottom Height**. In this case, I think that **-3 mm** should be more than enough. This way, we will machine more than needed, just in case our shoulder milling of the first placement was slightly misplaced.

The Passes tab

We can now move to the **Passes** tab, where we set the cutting depths.

Milling a hole is quite a demanding operation, the most critical part of which is the tool plunging inside the stock. A common strategy to reduce the workload on the inserts is to generate a helical toolpath while plunging. This type of toolpath looks similar to a spring or a coil.

We can use the **Passes** panel to properly define such a toolpath. To do this, we have two options, **Angle** or **Pitch**:

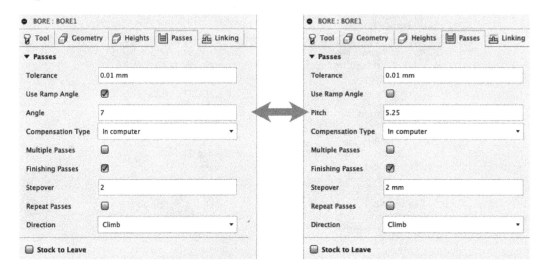

Figure 10.17: The Bore Passes tab

As you can see, we can decide to specify the helix angle by checking **Use Ramp Angle** and setting the **Angle** value (as shown on the left screenshot), or we can set the **Pitch** value (as shown on the right). Both methods ultimately result in the same generated toolpath.

I think it is a good idea to perform a roughing operation up to the bottom of the hole following such a helix. Once this roughing pass is completed, we can remove the remaining 2 mm for the final pass. Therefore, we shall tick the **Finishing Passes** box and set a **Stepover** of **2 mm**.

It is now time to set the parameters for the helix; however, there are a couple of hidden problems.

The pitch of the coil is equal to the axial depth of the cut we already calculated via the equations. The pitch and diameter are already enough to set a helix path; therefore, most of the time, we may forget to check the angle of the helix, and this is a dangerous mistake!

Here comes the catch. We must be careful since every tool has a maximum angle at which it can plunge into a part following a helical path. For the tool we picked, the limit is 7° (you can find this value in the URL previously provided in the *Implementing a roughing operation* section).

Therefore, we must be sure that the axial cutting depth that we specify doesn't lead to exceeding this angular limit! The axial depth of cut suggested by CoroPlus is 5.25 mm, but we must check whether such a pitch also respects the limit of 7° on the helix angle.

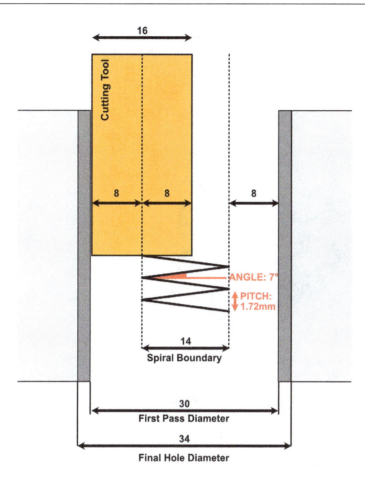

Figure 10.18: Helix angle and pitch

As you can see in the diagram, the hole to be machined is **34** mm wide; however, we decided to leave 2 millimeters all around for the final pass. Therefore, the hole has to be considered to be **30** mm wide.

Also, since the tool has a cutting diameter of 16 mm, we have to remove the radius from both sides of the hole, making the spiral boundary **14**.

Given these geometry constraints, a helix angle of 7° leads to a pitch of 1.72 mm, which is way below the calculated cutting depth of 5.25 mm. This means that instead of using the pitch parameter set to 5.25 mm, we should use the angle value of 7° leading to an axial depth of cut of just 1.72 mm.

Now that we have completed every roughing operation, we can check the results with a simulation:

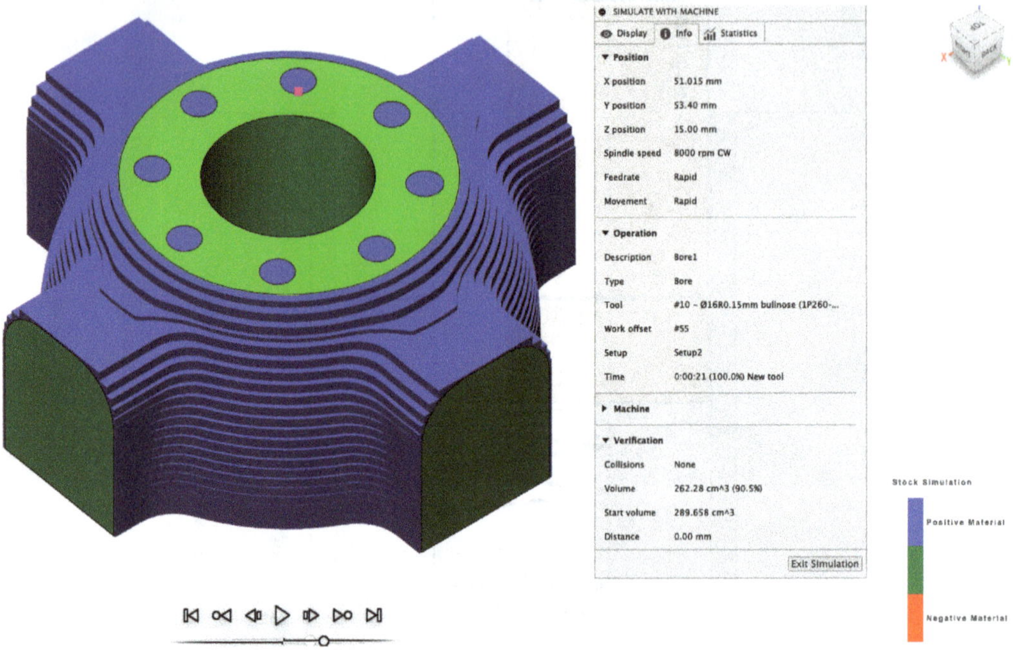

Figure 10.19: Simulation results

The simulation shows what we expected. Here is what we can find from the picture:

- No collisions were found
- The hole surface is green; therefore, we have no material left inside the hole, and it should be ready for threading
- The outer surface shows positive material due to the stock left on the shape with shoulder milling thick passes

Now we can finally say that we have completed the roughing of the part. There are just two more operations left: thread milling and profile finishing. We must change the cutting tool for both operations, so we can freely choose which one is to be performed first, but for this book, I would finish the outer area first.

Finishing the part using a morphed spiral

As we saw in *Figure 10.5*, there are several options for finishing, some of which have many traits in common and are sometimes even interchangeable. The idea behind them all is to implement multiple cutting passes very close to each other. The smaller the distance between these cutting passes, the better the overall surface finish.

For our example, I think that one of the best operations we can pick is **Morphed Spiral**. This command creates an adaptive path all around the part and is capable of machining complex shapes with steep surfaces.

As usual, before launching the command, we have to choose the right tool to use. When finishing a complex 3D contour (like in our example), a ball nose end mill is one of the best possible options. However, this type of tool would be a bad choice for us, given our shape and our three-axis machine. The reason is pretty simple: our part has flat surfaces!

As you can see in the following diagram, on the left, we have a ball nose tool approaching the flat surface along its normal direction. This is a huge problem since, on the contact point, the cutting radius is zero; therefore, the cutting speed is also zero. This means that in that area, the tool won't really be able to machine the stock properly.

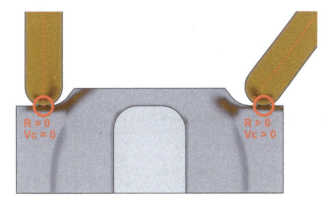

Figure 10.20: Ball nose issues

On the right, we can see a multi-axis machine approaching the flat surface at an angle. Tilting the tool completely fixes the problem since the contact point is not on the rotation axis anymore; therefore, the cutting speed is higher than zero, resulting in an easy milling operation!

We are not machining the part on a multi-axis machine; therefore, we simply cannot use a ball nose tool. However, we can rely on tools whose shape is not as entirely round as a ball nose tool, such as a flat-end mill.

The tool I decided to go for is somewhere in the middle: **RA215.26-4050HAL45L 1620.** It is a flat-end mill with a cutting diameter of 15.875 mm and quite a large corner radius since it is 3.175 mm.

You can find more details at the following link: `https://www.sandvik.coromant.com/en-us/products/pages/productdetails.aspx?c=RA215.26-4050HAL45L%20 1620&unitsystem=Metric.`

Now that we have picked our tool, we can finally launch the **Morphed Spiral** command and work our way through the tabs.

The Tool tab

First up is the **Tool** tab:

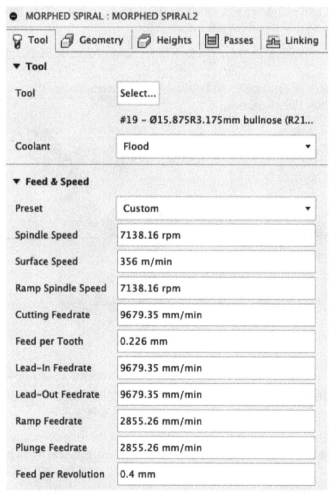

Figure 10.21: The Morphed Spiral Tool tab

The **Spindle Speed** we are going to use is **7138.16 rpm**, which leads to the **Surface Speed** being **356 m/min**. **Feed per Tooth** is **0.226 mm**; therefore, the **Cutting Feedrate** is **9679.35 mm/min**.

As you can see, these values are very high, particularly if we compare them to previous operations. However, we don't have to be too scared since the tool is not going to remove much material during the process; there is only a thin layer of material to get rid of. Furthermore, using such a fast set of parameters is actually desired since we have to reduce any rubbing against the already machined surfaces.

Anyhow, please keep in mind that this is our first approach to the part; if, when machining the first component, we find that the values are too extreme, we can always reduce them (hopefully before damaging the part or the tool).

The Geometry tab

The **Geometry** tab is pretty straightforward, but there is something we have to set. There is a typical problem, in fact; we have to specify which faces have to be considered for the operation. By default, Fusion 360 will try to machine every face with a layer of material yet to be removed. In our case, however, we just want to machine the outer area of the part and leave the central hole for the next operations.

Therefore, a typical fix is to create a surface cap to block the tool:

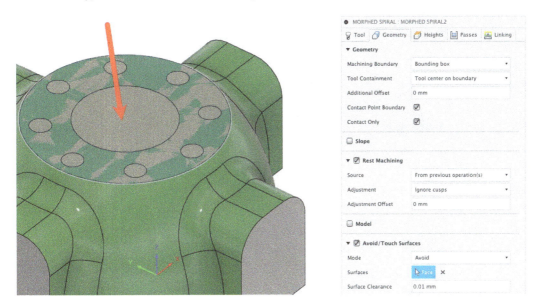

Figure 10.22: Surface cap to prevent unwanted machining

As you can see from the screenshot, I created a surface (using the **DESIGN** environment of Fusion 360) to cover the center of the part. This way, using the **Avoid/Touch Surfaces** option, I can be sure that the tool is not going to machine the center. This is a typical dirty way of fixing a toolpath that is too intrusive!

Other than that, you can leave the default parameters.

The Heights panel

The **Heights** panel has no hidden troubles to overcome. This time we can use the default settings, but we should increase the **Bottom Height Offset** a bit:

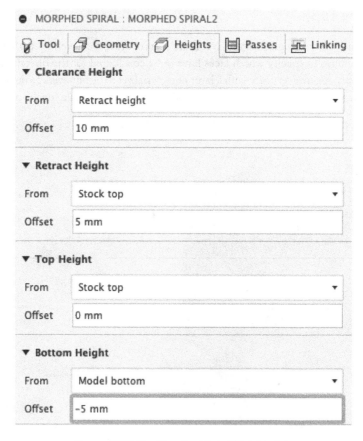

Figure 10.23: The Morphed Spiral Heights tab

Adding an offset to the bottom of the model is always a good idea (as long as we don't collide with the holding fixture) since we want to be sure that we remove the entirety of the material left from previous operations. Here, we will set the **Offset** to **-5 mm**.

The Passes tab

There are a few interesting parameters to note in the **Passes** tab, which we will look at here:

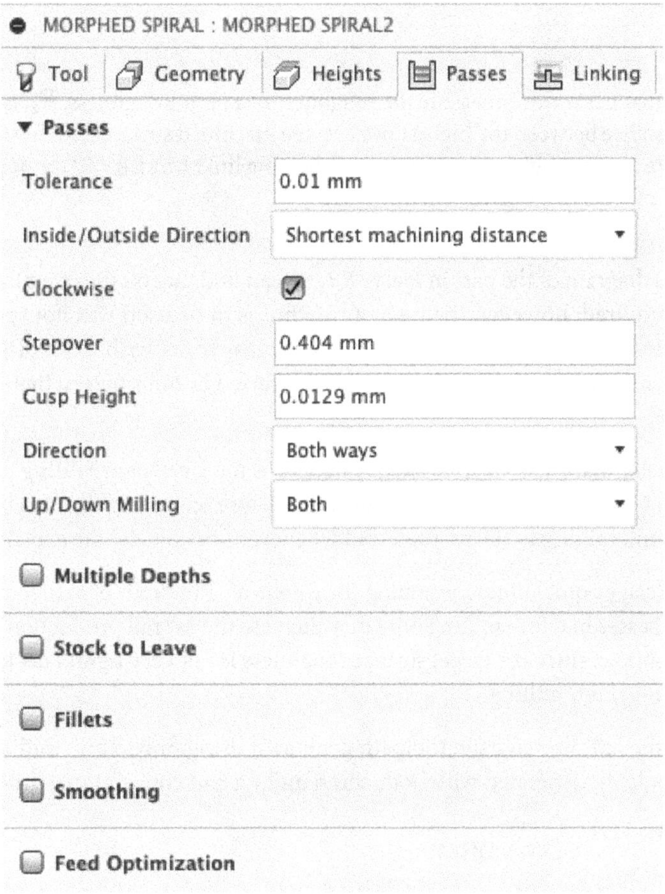

Figure 10.24: The Morphed Spiral Passes tab

Since the **Morphed Spiral** command creates a complex 3D path spiraling above the part, we can decide how to manage the generated toolpath; we have a few options to control it:

- **Inside/Outside Direction**: This controls whether the tool machines from the center outward or vice versa. Usually, the best idea is to leave the **Shortest machining distance** default setting, which should result in the fastest machining time.

- **Clockwise**: This is an option that will generate a toolpath following a clockwise or a counterclockwise direction of movement. Be careful: choosing one or the other may lead to larger areas machined in conventional milling rather than climb milling (it mostly depends on the geometry).

- **Stepover** and **Cusp Height**: These are strictly connected, so changing one will affect the other. **Stepover** measures how close the different cutting passes will be generated, while **Cusp Height** is the resulting roughness of the surface (R_z).

> **Note**
>
> There are several distinct ways to measure the roughness of a part. Roughness, R_z, is expressed in microns as the distance between the highest peak on the machined surface and the deepest valley. You can read more about roughness values at the following link: https://en.wikipedia.org/wiki/Surface_roughness.

If we look back at the diagram of the part in *Figure 8.1*, we can find that on the outer faces, a roughness of 12.9 microns is required; however, the type of roughness to be used was not specified. We will suppose it is expressed as roughness, R_z, since it is quite easy to set with this finishing command. In **Cusp Height**, we simply type the roughness value converted in millimeters; therefore, we enter a value of **0.0129 mm**.

An important parameter that you may have already noticed is the **Up/Down Milling** parameter, which this time is not set to **Climb** milling only, but set to **Both**. Before screaming that this is a huge mistake since we said that climb milling is always preferred whenever possible, let me explain my decision.

Finishing operations can be quite time-consuming since there are hundreds of cutting passes. Choosing to perform all these passes in climb milling only may increase the overall production time quite a lot; therefore, for this example, since the target surface roughness is not very tight, I decided to try to use both conventional and climb milling.

In *Figure 10.25*, on the left, we have the toolpath generated using only climb milling, while on the right, we have the toolpath generated with both climb milling and conventional milling:

Figure 10.25: Climb milling versus climb milling and conventional milling

As we can spot at first glance, the toolpath generated on the right has almost no travel movements compared to the example on the left. Also, the overall machining time is reduced by almost 30%, which is quite a big deal indeed; therefore, I think a less-than-optimal milling approach is worth a try at least!

Now that we have set everything needed for the operation, we can have a look at the simulation results:

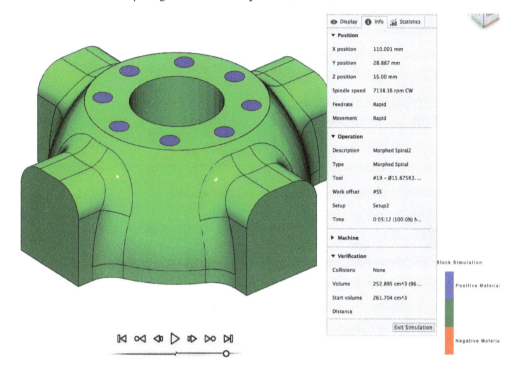

Figure 10.26: Morphed spiral simulation

So far, so good! The part looks amazing, with no extra material left on the surfaces, and the simulation found no collisions!

We can now move to the final operation, which is thread milling. It will be our last milling example and also, be aware, one of the most complex.

Thread milling

The last operation we have left to complete the part is threading the central hole with an M36x4 profile.

During the previous operation (in the *Milling a hole* section), we already machined the pilot hole, which has a diameter of 31.97 mm.

We should also already know the theory behind threads since we already covered a similar operation in *Chapter 5*; however, this time we'll cover the process in a bit more detail and try to explain a few potential issues we didn't mention before.

Thread geometry

In order to machine threads, we have to know them! To find all the information needed for our thread, we can discover ISO thread tables online.

When searching the web for thread tables, we have to make sure that we are looking at the right type of thread. Internal and external threads have different dimensions, and confusing one for the other will result in a wrong thread profile.

In the following table, we can find the data for an M36x4 internal thread; we will need these values for the calculations ahead:

Size	Pitch	Pitch Diameter	Major Diameter	Minor Diameter
M36x4	4	33.402	36	31.670

Figure 10.27: M36x4 internal thread specs

Please note that when looking at online threads tables, you may find that real tables are a bit more complex than the one provided here. In reality, on good-quality tables, there is always a certain tolerance on acceptable diameter dimensions; this will result in finding the maximum and the minimum value that every parameter can assume. Since this book is about CAM and not about milling threads, we will oversimplify this part, ignoring machining tolerances.

Before jumping to tool selection, we have to spend a couple of words on how thread milling is set using Fusion 360. The key idea is that after selecting the hole cylinder, we have to specify how much the tool should plunge into the part to machine the thread:

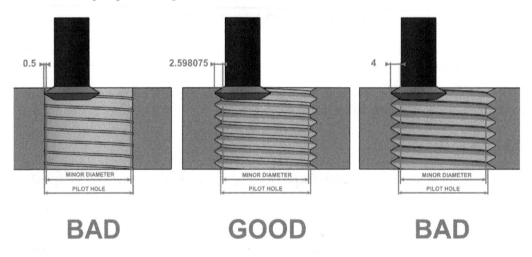

Figure 10.28: Good versus bad thread milling

As we can see from the diagram, it is important to have a proper thread that is up to the correct specifications; even minor errors in the thread geometry will result in a non-working part.

Long story short, we have to calculate how much the cutting tooth should extend beyond the limit of the minor diameter. To do this, we need to refer to the thread drawing already used for turning:

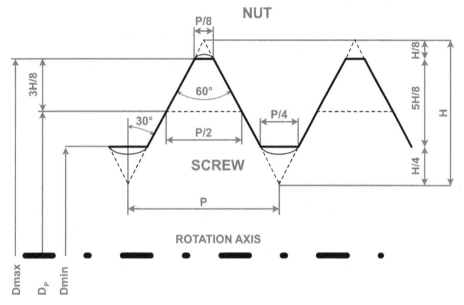

Figure 10.29: ISO thread specs

Here we can see several dimensions using letters such as P and H. P is, of course, the pitch, which in our case is 4, while H is the theoretical thread height, which can be calculated using the following formula:

$$H = \frac{\sqrt{3}}{2} * P = \frac{\sqrt{3}}{2} * 4\ mm = 3.464\ mm$$

As we can see from the diagram, the thread height for an internal thread can be expressed in two ways:

- The first formula uses H as its main parameters:

$$\frac{5}{8}H = \frac{5}{8} * 3.464\ mm = 2.165mm$$

- While the second formula is simpler if we already know all the thread specifications (which we do):

$$\frac{D_{max} - D_{min}}{2} = \frac{36 - 31.67}{2} = 2.165mm$$

This may sound like we solved the problem of finding the radial depth of cut for our cutting tool, but there are two critical points we need to pay attention to.

First of all, this value is the thread height for an internal thread, and it is measured from the minor diameter of the thread; as we can see in the thread dimensions table, an M36 is equal to 31.67 mm.

The problem is that the cylindrical face at the center of the part has the diameter of the pilot hole, which is 31.97 mm and is bigger than the minor diameter of the thread. Therefore, if we want to work precisely, we need to create an additional surface at the center of the cylinder corresponding to the minor diameter. Such a surface will then have to be selected as geometry for the thread milling command instead of the pilot hole.

Secondly, we have to keep in mind the tool shape and how its tip is modeled inside Fusion 360:

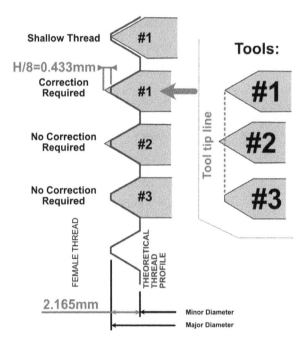

Figure 10.30: Tip offset

As we can see from the diagram, there are three tools: tool #1 and #2 have a sharp tip, while tool #3 has a flat end. The origin of tools #1 and #3 is placed on the tip itself while the origin of tool #2 has a backward offset.

Since every tool plunges inside the stock by the same value (**2.165 mm** for our example), and since this distance is measured from the tool origin, tool **#1** will machine a shallow thread; hence, it will require a correction factor (or offset) equal to **H/8**, which for the example is **0.433 mm**. Tools **#2** and **#3** will machine the thread properly since their origin is properly placed.

Don't be too scared of what we just discovered; it was simply to remark that we should always be careful with tool origin placement, especially with custom tools!

Picking the proper threading tool

Now that we have a clear idea of thread milling, we can move to tool selection and parameters. I decided to use the following cutting insert named **327R12-22 250VM-THM 1025**. You can find more information at the following link: `https://www.sandvik.coromant.com/en-gb/products/pages/productdetails.aspx?c=327R12-22%20250VM-THM%201025`.

Let's analyze the most important details of this type of tool to see whether it is the right tool for our job. First of all, we have to understand whether this tool is capable of milling a thread with the pitch we need.

The online datasheet shows that the tool can machine between 5 and 10 **threads per inch** (**TPI**). Converting threads per inch to threads per millimeter gives us a working range from 0.19 to 0.39 threads per millimeter (we just have to divide the TPI value by 25.4).

Now that we have the number of threads per millimeter, we need to find the pitch:

$$Pitch[mm] = \frac{1}{TP[mm]} \quad or \; Pitch[mm] = \frac{25.4}{TPI\,[in]}$$

Inserting the values inside the formula, we get the following results:

$$PitchMAX[mm] = \frac{25.4}{5[in]} = 5.08mm \quad PitchMIN[mm] = \frac{25.4}{10[in]} = 2.54mm$$

This tool can machine thread pitches in a range between 5.08 mm and 2.54 mm. Our thread has a pitch of 4 mm, so this is the right tool.

Let's calculate the theoretical thread height we need. The minor diameter of the thread is equal to 31.67mm, while the pilot hole is 31.97mm; therefore, their difference is 0.3mm, meaning there is 0.15mm on each side of the pilot hole. To this value we shall add our thread height, which is 2.598mm (6/8H).

Long story short, we need a tool with a theoretical thread height of at least 2.748mm. Therefore, the tool used in the example shall work properly for us.

Implementing thread milling in Fusion 360

We can now launch the **Thread** command, which we can find inside the **2D** milling commands dropdown:

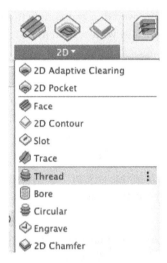

Figure 10.31: The Thread milling command

Let's take a look at the tabs.

The Tool tab

From the tool datasheet, we don't find any suggested parameters for aluminum alloys; therefore, we have to pick the parameters for steel. Those parameters will probably be quite conservative for aluminum milling; therefore, we may decide to increase them to maximize production later. As an alternative, we can always calculate the parameters ourselves!

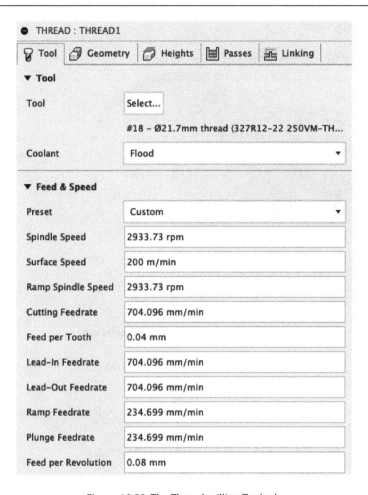

Figure 10.32: The Thread milling Tool tab

For our example, we will set **Surface Speed** to **200 m/min**, which leads to a **Spindle Speed** of **2933.73 rpm**, and a **Feed per Tooth** of **0.04 mm**, which leads to a **Cutting Feedrate** of **704.096 mm/min**.

The Geometry tab

Inside the **Geometry** tab, there is not much we have to do:

Figure 10.33: The Thread milling Geometry tab

Here, we simply have to select the cylindrical surface. Please note that, as already mentioned, we don't have to select the pilot hole (which has a diameter of 31.97 mm).

I think the simplest approach for a beginner is to design and select a surface whose diameter corresponds to the minor diameter for our thread.

In our case, an M36 internal thread has a minor diameter of 31.67 mm.

The Heights tab

We can now move to the next panel, which is called **Heights**:

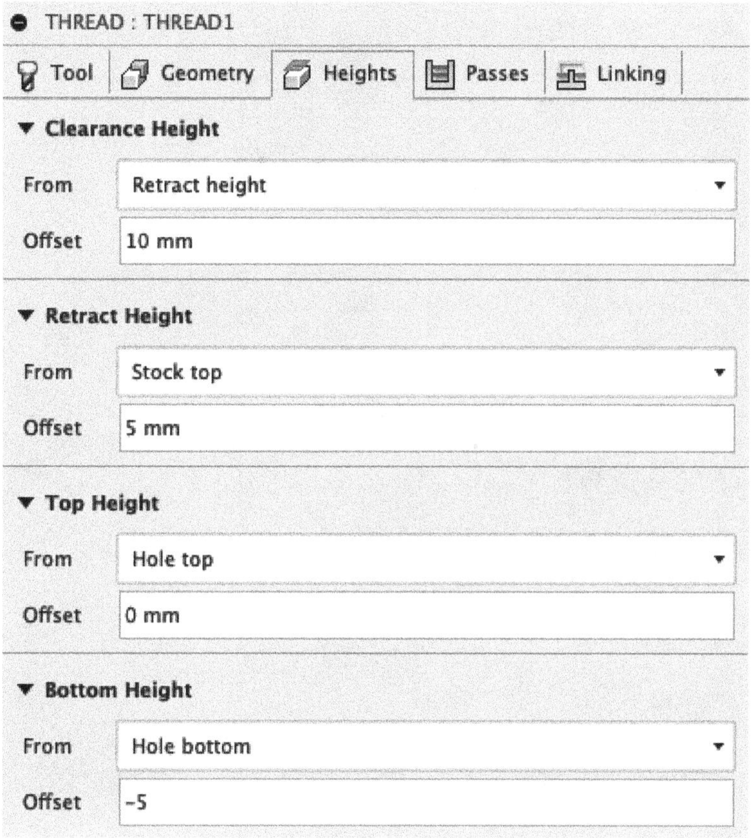

Figure 10.34: The Thread milling Heights tab

This window is no surprise for us either; much like we did for many other operations, we just want to add a bit of offset on the bottom of the thread to be sure to perfectly machine the entire length of the hole. For this example, we will simply add around **-5 mm**.

The Passes tab

The **Passes** tab, as usual, is a bit trickier:

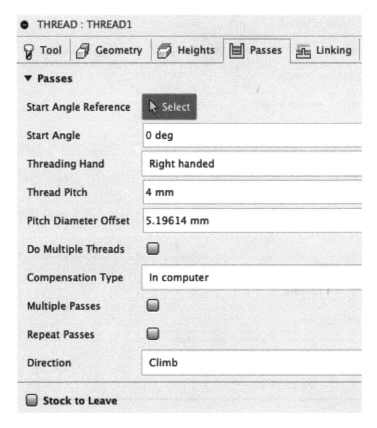

Figure 10.35: The Thread milling Passes tab

As you can see, there are several options already discussed time and time again, but others are new and a bit confusing. In the following list, we will review the new options available:

- **Start Angle Reference**: This parameter, combined with **Start Angle**, allows us to specify the angle at which the thread shall begin. Most of the time you don't need to specify this value since almost no one cares about thread angle.

- **Threading Hand**: With this option, we can specify whether to create a **Right handed** thread or a **Left handed** thread. Since this is thread milling and not tapping, we don't have to change the tool, as the threading mill can machine both right-handed and left-handed threads.

- **Thread Pitch**: This describes how much a thread is screwed in and out of a threaded hole in one revolution. In our case, the pitch is **4 mm**.

- **Pitch Diameter Offset**: This is the most difficult parameter to understand; basically, it is how much the cutting insert plunges inside the stock when machining the thread. It is measured from the selected cylindrical surface to the tool origin. In our case, we can calculate the thread height as 5/8H, which is equal to 2.165 mm. Alternatively, if you recall from the *Thread geometry* section, we can calculate the same value with the following formula:

$$\frac{D_{max} - D_{min}}{2} = \frac{36 - 31.67}{2} = 2.165mm$$

Since the tool we are using has the origin on the sharp tip, I shall add an extra 1/8H for a total of 2.598 mm. In addition, this value is related to the diameter, so we have to multiply the value by two, resulting in a value of **5.19614 mm**.

- **Do Multiple Threads**: This is a typical parameter for high-pitch lead screws; this is not the case for us, of course.

Now that we have completed this operation, we have reached the finish line; the part is now completed, and we can admire the result!

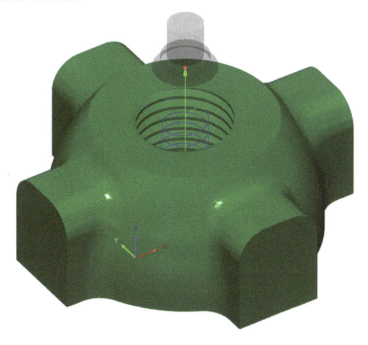

Figure 10.36: Second placement completed

Wow, what a result! We should be proud of it!

Summary

Congratulations, we have now finished the section on milling! I'm sure that we had a good time together creating our machining strategies from scratch.

In this chapter, we faced several advanced milling strategies: we used one of the best milling operations that Fusion 360 has to offer, which is **Adaptive Clearing**, then we found out why it is important to find a sweet spot between roughing passes and finishing passes, and how tweaking retract movements or milling direction can increase productivity.

After that, we took quite a deep dive into thread milling, where we tried to uncover every possible parameter needed for proper machining and we found out how to check whether the selected tool was up to the task.

This was probably the most difficult chapter to understand due to the geometries and the challenges proposed; however, I promise you that the hardest part is now over.

It is now time to leave milling behind and finally move on to a new manufacturing technology. In the next chapter, we are about to discover laser cutting!

Part 3 –
Laser Cutting Using
Fusion 360

In this part, we will discover the basic concepts behind laser cutting, a fascinating technology with so much potential. Not only will we discover how a laser cutting process is managed via CAM, but we will also take a deep dive into nesting and its value for cost optimizations. Ultimately, we will create a cutting toolpath ready to be exported to our laser cutting machine.

This part includes the following chapters:

- *Chapter 11, Getting Started with Laser Cutting*
- *Chapter 12, Nesting Parts for Laser Cutting*
- *Chapter 13, Creating Our First Laser Cutting Operation*

11

Getting Started with Laser Cutting

In this chapter, we are going to introduce laser cutting, including its strengths and drawbacks and the potential it can give you.

Fusion 360 has quite a limited set of commands and tools dedicated to laser cutting and is not exactly the best solution when it comes to laser cutting. However, getting a general idea of a new technology is still very important since it will help us to expand our knowledge beyond milling and turning.

Therefore, the goal of this chapter (and the following ones on laser cutting) is to provide several hints about laser cutting. These chapters are not really aimed at industrial production. We will take a discursive approach to the subject, so we won't be diving too much into parameters or equations.

In this chapter, we will cover the following topics:

- Introducing lasers
- How does a laser cut?
- Reviewing the pros and cons of laser cutting

Technical requirements

We have now changed the production technology, moving from milling to laser cutting; therefore, this chapter is a fresh restart – there are no prior requirements.

Introducing lasers

Lasers, known to humankind since the ancient Egyptians, proved themselves to be valuable time and time again during the construction of the pyramids – at least this is something a ufologist may suggest to us.

Jokes aside, in reality, lasers date back to 1916 when Albert Einstein first theorized their behavior. As you may expect from such an important physicist, grasping the fundamental nature of lasers is rather complex and would require us to be quantum physicists. That kind of understanding is outside the scope of this book, so instead, we will simplify things quite a bit.

Long story short, a laser is an energetic beam of light that is coherent and focused on a very tiny spot. I know that this may sound weird to most, but light is far more complex than most of us would imagine.

Light is an **electromagnetic (EM)** wave that we can see with our eyes. Saying that, not every EM wave is visible to the eye; it mostly depends on its wavelength, which is a property of every EM wave.

In the following diagram, we can see a large chunk of the **electromagnetic spectrum** ranging from 1 picometer (1 pm is equal to 0,000000000001 meters) on the left up to 1 meter on the right. Then, we can find vertical lines that divide the spectrum into subcategories of EM waves:

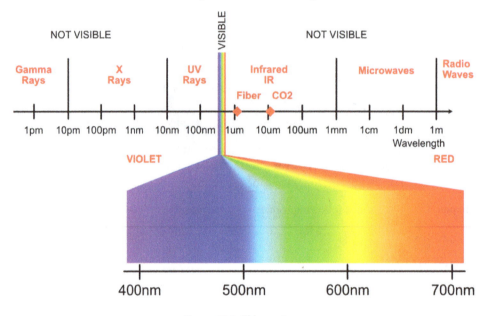

Figure 11.1: EM spectrum

Taking a couple of examples, here we can see that waves with a wavelength over **1m** are **Radio Waves**, and waves between **1m** and **1mm** are **Microwaves**.

As you may have also noticed, inside the visible range of the EM spectrum, different colors are associated with different wavelengths. For example, light rays with a wavelength of 740 **nanometers (nm)** are seen as red, while light rays with a wavelength of 380 nm are violet.

Furthermore, EM radiations with wavelengths bigger than 740 nm or smaller than 380 nm are not visible to our eyes. For example, we cannot see X-rays or radio waves because their wavelength is beyond the limit of our eyes, but of course, they are real nonetheless.

This prelude on wavelengths is needed to highlight a very basic concept: despite what we may have learned from Hollywood movies, lasers may be both invisible and deadly!

Included in *Figure 11.1* are two dots on the **infrared** (**IR**) range, which mark the two most common types of cutting laser wavelengths for the manufacturing industry: **CO2** lasers and **Fiber** lasers. Both of these lasers are completely invisible to our eyes and yet can instantly destroy our retina.

> **Note**
> Always wear protective goggles when working with lasers and read the safety instructions when operating close to a laser machine.

Something else to note is that not every type of laser can cut every material. It mostly depends on the behavior of the material we need to cut, how it reacts to the light beam, and how much of the incoming radiation is absorbed, refracted, or reflected.

For example, a CO_2 laser can cut cardboard and wood effortlessly, while a fiber laser cannot. On the other hand, a fiber laser can cut very shiny materials such as copper and aluminum much better than a CO_2 laser can. Choosing one type of laser over another mostly depends on your needs.

In this book, we will focus on sheet metal cutting from the point of view of welded steel components. For this type of manufacturing, fiber lasers are absolutely the leading technology; therefore, we won't deal with other types of lasers anymore.

Before we do any laser cutting, in the next section, we will find out how a beam of light can cut a solid, tough chunk of metal!

How does a laser cut?

In recent years, the power of fiber laser sources has exploded, and every year, new records are achieved. A few years ago, a laser source with a power of 6 **kilowatts** (**KW**) would have been a dream come true for most producers; however, nowadays, it is typical to find cutting machines with a 15 KW source, with some producers even starting to experiment with sources up to 50 KW.

To wrap our head around these numbers, let's imagine my motorbike, which has a power of 50 KW; at full throttle, it can bring its mass (190 Kg) and my mass (75 Kg) all the way up to more than 200 km/h. Now, imagine all that power focused on a tiny spot with an area of less than a square millimeter. As you may guess, focusing all this energy on such a small surface leads to temperatures high enough for the metal to melt (or vaporize).

Melting the metal, however, is not enough to cut it properly: we need to remove the melted metal and clean the cut width; otherwise, once the laser moves, the melted part will solidify again!

In the following diagram, we get an idea of how a laser-cutting machine works:

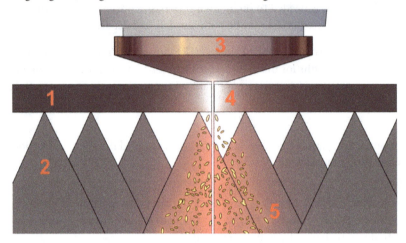

Figure 11.2: Laser cutting

As we can see, the metal sheet (**1**) is laid on the bed of teeth (**2**), and the laser nozzle (**3**) is very close to the surface it is cutting. In the area where the laser beam is focused (**4**), the metal reaches very high temperatures and melts down. All the melted droplets are evacuated from the bottom (**5**) thanks to high-pressure gases that clean the cutting area. The nature of such gases can vary according to the material being cut, but the most common gases used are **oxygen (O2)** and **nitrogen (N2)**.

Is this all we need to know to set up a cutting process? Of course not; there is the shape of the laser nozzle to choose from, the type of gas, the power and pulses of the laser to set, and so on. There are lots of variables in laser cutting, but luckily, we don't really have to understand or calculate them all.

Almost every professional laser cutting machine has its own CAM software and a material database already tweaked for the best performance with that particular machine. This way, we can simply input the material type, thickness, and 2D profile of the material that we want to cut and let the post-processor manage the cutting parameters.

There are several software suites specific to laser cutting, one of the most famous solutions being Lantek (which you can find more about here: www.lantek.com).

Fusion 360 is a bit different – it is not focused on a single specific task such as laser cutting; rather, it is intended as a general software suite to bring our projects from concept to production. For this reason, sometimes it is not as complete as other alternatives.

Therefore, currently, there is not a great number of laser operators that use Fusion 360 for laser cutting since there are far better solutions!

> **Note**
>
> As we are about to discover, most of the parts cut with laser machines consist of 2D profiles; therefore, the most common file type used for laser cutting is `.dxf`. For this reason, it is not rare to find designers still using 2D software such as AutoCAD, CorelDraw, or even Inkscape.

Now that we have an idea of the cutting process, we can move on to discover the main features of laser cutting.

Reviewing the pros and cons of laser cutting

Cutting metal sheets with a laser is a valuable option for a wide range of applications, from rough welded parts for building sites to artistic decorations that can be found in finely furnished homes. However, as you may expect, no manufacturing technology is free of limitations or drawbacks.

Let's review the pros and cons of this process.

Advantages of laser cutting

Let's start by discussing why laser cutting is such a great technology.

Extreme productivity

In other technologies (such as milling or turning, for example), the tool powerfully plunges into the stock to remove the chip. This type of strong interaction with the part needs a very stiff CNC structure that doesn't bend too much and a strong stock holding fixture to counteract such a force.

As we found out, laser cutting is achieved without touching the part with any component of our machine. This means that as long as the laser source can generate enough power to melt the material, we can cut at practically whatever speed we see fit, achieving impressive speeds and accelerations that would be impossible for any milling CNC center. Nowadays, there are several laser machines that move up to 300 m/min with accelerations of 2-3 g, numbers impossible for any other manufacturing process.

> **Note**
>
> **Acceleration** is a value that expresses how fast the speed is changing. It is measured in **meters over seconds squared** (m/s^2). A very common practice is to express acceleration as a factor of g (which is the earth's gravity acceleration and has a value of 9.81 m/s^2).
>
> Therefore an acceleration of 3 g equals a value of 29.43 m/s^2.

No chip formation and waste friendly

It may sound obvious, but laser cutting creates no chip! Since the width of the laser cut is very small, around 0.2–0.3 mm, for all practical purposes, we can consider that no material is actually removed. The only wasted material is due to the areas not being optimized on the sheet **nesting**.

Nesting is a procedure aimed at optimizing the placement of parts onto a metal sheet. In the following screenshot, we can get a clearer idea of what nesting is all about:

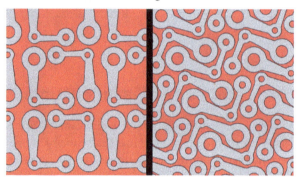

Figure 11.3: Nesting

As we can see on the left, we have quite a conservative nesting with a lot of material not used for the parts, while on the right, there is a tightly packed nesting pattern that will lead to low amounts of wasted material.

We will explore nesting in more detail in *Chapter 12*.

Highly flexible

Laser-cutting machines can cut a whole range of materials and thicknesses.

Until a few years ago, a laser could cut only thin metal sheets, and the only option for thicker plates was another cutting process called plasma cutting.

Today, thanks to powerful fiber sources, laser cutting can easily cut very thick chunks of metal, gradually overtaking plasma cutting (and it is not hard to imagine that it will replace it entirely sooner or later).

Not only has the expansion of fiber technology drastically increased cutting powers but it has also reduced the costs of production and maintenance, meaning nowadays, several companies offer their laser machines at compelling costs.

Drawbacks of laser cutting

It is now time to jump into the drawbacks of laser technology, though several points may be unappealing to you.

Heat generation

Since lasers cut the material by melting or vaporizing the spot where they are pointed, this heat may alter the metal's crystalline structure.

To put things simply, if you have ever seen a sword maker creating a sword, you may have noticed that once the sword is very hot and red, it is rapidly submerged in a barrel of cold water. This process is called **quenching**.

The fast cooling of the metal forces its internal molecules to arrange in a lattice that enhances the mechanical properties of the sword, which becomes stiff and hard but also quite fragile. A later controlled heating of the sword will give the blade its ultimate properties, a sweet point between stiffness and toughness.

With laser cutting, uncontrolled heat generation and cooling is a problem that may introduce weak spots or distortions on the part; however, a proper material database and the right management of cutting parameters can greatly mitigate this scenario.

Calamine

Another problem related to heat occurs when cutting certain types of steel alloys (using O_2 as cutting gas). In this scenario, the cut surface gets covered in a dark layer of **calamine**. Calamine is an oxide created as a subproduct of the highly **exothermic** reaction between oxygen and steel.

> Note
> An exothermic reaction is a chemical reaction that generates a large amount of heat.

Other than looking bad, this layer of calamine is not tightly bound to the steel and can be peeled off very easily. Painting (or welding) on a very thin and fragile layer of material is not a great idea as the final result would be compromised: the paint would fall off immediately, and the surface would rust in no time!

This layer of unwanted material has to be removed mechanically, using abrasives, or chemically via pickling. Luckily these are easy solutions, though still inconvenient overall.

Sharp edges and surface finish

Laser cutting tends to create sharp edges on the cut area, so handling thin metal sheets without gloves will probably end with a wound on our hands. This is also true for thicker materials. In general, as a rule of thumb, the thicker the material, the worse the surface finish will be:

Figure 11.4: Sharp edges

The image shows a 30 mm plate of aluminum cut with a high-power laser beam. As we can see at the top of the picture, the laser nozzle was closer to the surface, so the finish is quite smooth and homogeneous. But moving toward the bottom of the plate, the roughness gets higher and higher, giving it a distinct, sharp, saw-like profile.

Only works with 2D profiles

This may be quite obvious to some, but laser cutting doesn't let us cut 3D shapes. Laser cutting is like having a blank sheet of paper and a pair of scissors: we cannot cut at variable depths, and we cannot cut using multiple placements. This is true for 99.99% of the laser machines out there.

However, what about the other 0.01%? There are some rare models of laser machines capable of tilting the head for four- or five-axis cutting. However, don't get too excited, as this option is mostly used to create chamfers for welding or tube cutting.

As a rule of thumb, we just have to remember that the shape complexity and accuracy obtainable with laser cutting is way lower than milling.

Overall, laser cutting is an interesting technology with extreme productivity and excellent value for money, but with some limitations we should always consider. Saying that, it's a good tool to have in our hands.

Summary

And that's the end of the chapter. Here, we introduced lasers, looked at how a laser machine works, and started to understand the underlying complexities and considerations that proper CAM software manages to cut parts with a laser.

Despite not being able to call ourselves experts on the subject after these few pages, I hope that we now have a general idea of what is going on with laser cutting and when it may be an interesting manufacturing solution to opt for.

In the next chapter, we will continue our laser-cutting introduction by exploring how to create a proper nesting for our parts!

12

Nesting Parts for Laser Cutting

This chapter will cover the topic of nesting optimization as deeply as possible without resulting in being too difficult for a novice.

As we are about to discover, learning how to manage nesting for a cutting process is essential to reduce production time and costs. Proper production optimization for large batches is quite a complex topic to cover since it consists of manufacturing processes highly intertwined with logistics and warehouse management. Therefore, we must simplify the subject and will focus on the manufacturing point of view only.

The goal of this chapter is to introduce all the tools that Fusion 360 offers for a proper nesting setup and the other options available. In addition to the basic software included in Fusion 360, we are also about to explore a valuable extension (called Nesting and Fabrication) with several additional features.

In this chapter, we will cover the following topics:

- Presenting the example model
- Understanding nesting optimization
- Creating a nesting with Fusion 360

Technical requirements

To be able to properly follow this chapter, make sure that you have read *Chapter 11*.

The only special requirement to follow this chapter is to have a working license for the Nesting and Fabrication extension for Fusion 360. It is a subscription piece of software that can be obtained using the following link: `https://www.autodesk.com/products/fusion-360/nesting-fabrication-extension?term=1-YEAR&tab=subscription`.

If you just want to activate a trial license, you can visit the following link and follow the on-screen instructions: `https://www.autodesk.com/products/fusion-360/extensions/trial-intake`.

However, this plugin is not mandatory since the key concepts and ideas are still valuable even without such an expansion.

Presenting the example model

As simple as it may sound, in order to create a nesting study, we should first start with a kit of parts to be nested. We will work with a welded structure that I created for that purpose:

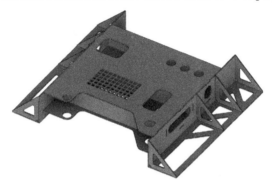

Figure 12.1: Example part

As we can see from the diagram, it is a rather complex part with several different components. It is composed of 3 mm mild steel sheets, which are assembled and welded together.

There are a total of five different parts in the assembly that are repeated multiple times:

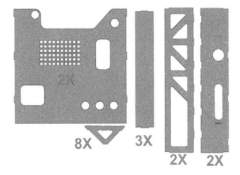

Figure 12.2: Example part instances

To create a kit of parts for the welded structure, we need to cut five different components in different quantities. The larger part has to be cut twice, and the smaller one has to be cut eight times, and so on.

Now that we have a clear understanding of the example part and its internal components, we need to start thinking about nesting optimization. As we discovered in *Chapter 11*, optimizing nesting is the best way to reduce production time and waste material. In the next section, we will study this important process.

Understanding nesting optimization

Nesting is all about placing as many parts as possible on the stock metal sheet we are about to cut. It's a bit like aiming for a Tetris record where the shape of the brick has to conform to all the others to pack everything as close as possible.

However, we don't really have to manually orient the components on the boundaries of the stock metal sheet, as there are now automatic processes that optimize part placements in just one click.

Saying that, there are two important factors to consider that are strictly interconnected with nesting: sheet format and batch volume. Let's look at those now.

Sheet format

Sheet format is the easiest variable to optimize – we simply need to consider the size of the metal sheet we want to cut.

As we all know, paper sheets are sold in standard formats such as A4 and A3 and so on. Metal sheets work the same way; there are several standard formats, some bigger and some smaller. From an industrial point of view, most professional laser machines have a format of 3000x1500 millimeters. However, we can also pick any smaller sheet size; as long as the raw metal sheet is bigger than the bigger part we need to cut, we are good to go.

For our example, we have a smaller machine capable of handling panels with a maximum dimension of 1000x500 millimeters, and this will be our format. Given these dimensions, we may start playing Tetris and optimize the placement of the parts in such a format, like so:

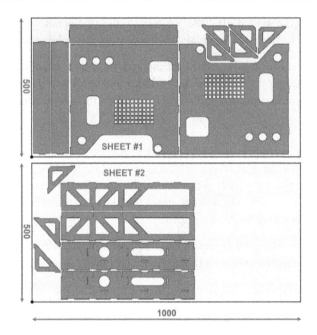

Figure 12.3: Nesting example #1

As you can see, even the best Tetris player won't be able to put every part of our welded structure onto a single metal sheet; however, as we have done, we can create one set of parts (one kit) using two metal sheets.

So, the problem is solved: the minimum number of metal sheets we will use to create our part is two. If we need two welded assemblies, we have to cut four metal sheets, if we need to produce three, we have to cut six sheets, and so on.

But not so fast; there's a catch.

Most of the time, a production order is not for a single welded assembly, probably we have to produce more than one; for example, we may have to produce 10 of them or 100. Therefore, instead of thinking in terms of a single kit of parts needed to produce a single welded assembly, we have to start optimizing the entire production lot (also called batch volume).

Batch volume

Optimizing nesting for the production of a single kit of parts for our welded structure is, for sure, a really flexible solution, but it may not be fully optimized for large batches.

As you may have gathered yourself from *Figure 12.3*, there is quite a large area of the sheet not used – if we keep using the same nesting even for larger production batches, that area will always result in material waste.

A much better solution is to create nesting that is optimized for the production batch; this way, the placement will have much more room to play with. Fusion 360, instead of trying to fit all the parts contained in a single kit in the least amount of sheets, will try to fit all the parts for all the kits in the least amount of sheets. It is a subtle difference that makes a huge optimization improvement!

Let's suppose we actually need to produce three kits of parts to assemble three welded structures:

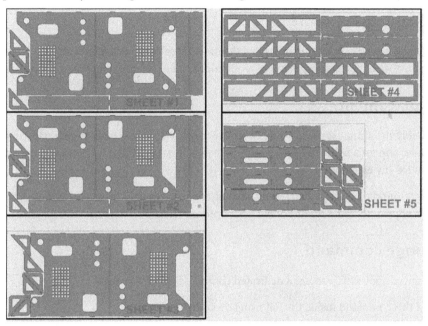

Figure 12.4: Nesting example #2

As we can see here, the overall number of needed sheets is reduced: optimizing a nesting for a bigger batch assures an overall smaller number of metal sheets used.

This time, in order to produce three kits to weld, we only needed five metal sheets, compared to the previous nesting example (displayed in *Figure 12.3*), which needed six sheets! Not only does this improvement mean less scrap material, but it also means we get higher productivity since we repeat the loading and unloading process fewer times.

For smaller batches, such a gain is negligible, but once the batch is in the hundreds, the advantages are clear. The only drawback to batch-optimized nesting is that if, for any reason, the batch number changes, we may need to create another nesting from scratch.

Now that we have a good idea of how to create a proper nesting, let's find out how to create it with Fusion 360.

Creating a nesting with Fusion 360

There are three different approaches for creating a nesting in Fusion 360, each with different potential:

- Manual placement
- The Arrange command
- The Nesting and Fabrication extension

Let's review them now. Please keep in mind that the most complete toolset for nesting is not always the best way to go; if we have to manage simple productions, we may want to opt for a simpler solution.

Manual placement

This is probably the simplest yet most flexible way to go. We simply have to manually orient the parts onto a certain area using the **Move** tool in Fusion 360, much like we do inside the **DESIGN** environment for any assembly.

As simple as it sounds, this approach lets us position the parts exactly as we see them fit; however, it is the longest approach, and it doesn't allow great optimization.

The Arrange command

A more advanced approach is to use a dedicated command called **Arrange**.

You can find this command inside the following drop-down menu:

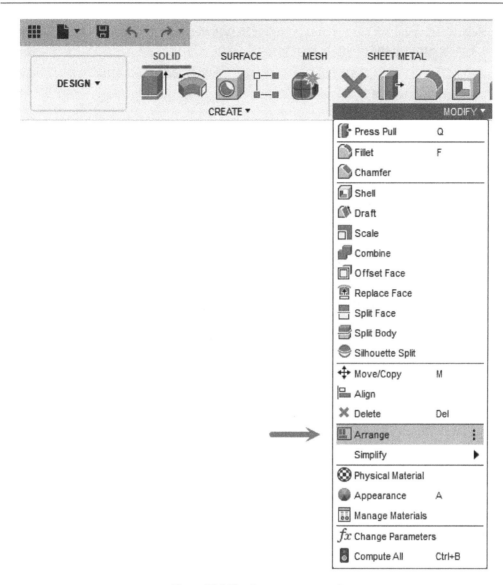

Figure 12.5: The Arrange command

If you cannot find the command or if it is grayed out inside the drop-down menu, you probably have to enable it inside the options of Fusion 360.

To enable this command, we must go inside Fusion 360's **Preferences**, then the **Design** panel, and click the **Enable Arrange and Simplify tools** option:

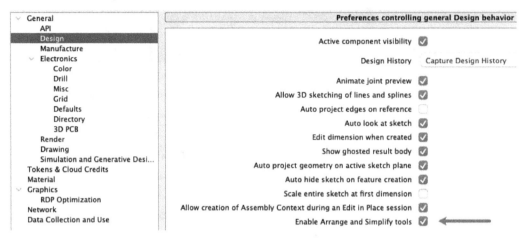

Figure 12.6: Enabling the Arrange command

Then restart Fusion 360 to make sure the **Arrange** command is available.

A quick refresh of components

The **Arrange** command is capable of arranging the components of our assembly onto a flat surface. Be careful: the word *component* is essential here, trying to use the **Arrange** command with *bodies* won't work at all.

Before reviewing how to create a nesting using the **Arrange** command, I think it is important to remind ourselves of the difference between components and bodies.

Fusion 360 has three main types of elements inside the geometry hierarchy: **bodies**, **components**, and **assemblies**. Since this book is not about the **DESIGN** environment of Fusion 360, we won't dig too much into the differences between these elements, but let's just say that bodies are basic geometries such as solids and surfaces, components are groups of one or more bodies and are physical objects, and assemblies are groups of components or sub-assemblies.

To refresh your mind, here is a typical structure for any Fusion 360 file:

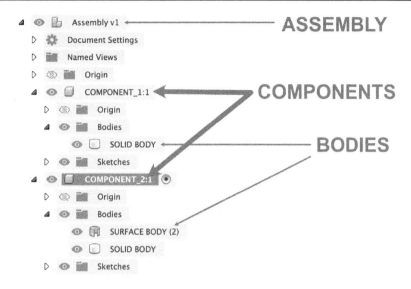

Figure 12.7: The assembly structure

As we can see, there is an assembly called **Assembly v1** that contains two components called **COMPONENT_1:1** and **COMPONENT_2:1**.

Inside **COMPONENT_1:1**, we have a single solid body, while inside **COMPONENT_2:1**, we can find a surface body and a solid body.

Sometimes we may have forgotten about this difference and created a large group of bodies that should be components instead. This is not an issue, as we can convert them to components by right-clicking on the body and selecting **Create Components from Bodies**:

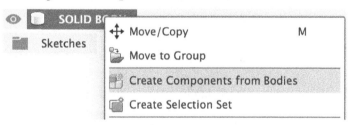

Figure 12.8: Create Components from Bodies

> **Note**
> Please be careful; at the time of writing, sheet metal bodies cannot be converted to components!

This concludes the digression into Fusion 360 geometry hierarchy.

Using the Arrange command

Now that we know the difference between bodies and components, we can keep going with the **Arrange** command. Let's launch it using *Figure 12.5* as a reference. You will then see the panel shown in *Figure 12.9*:

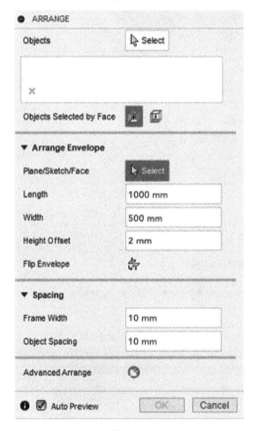

Figure 12.9: The Arrange panel

The panel for this command is quite limited, but let's take a better look at it:

- **Objects**: This selection tool lets us specify the objects we want to arrange.
- **Plane/Sketch/Face**: This selection tool lets us specify the geometry to use as a plane reference. The components to be nested will be oriented on this plane.
- **Length/Width**: With these values, we can set the maximum dimensions of the nesting result.

- **Frame Width**: With this value, we can set a constant offset from the outer contour of our metal sheet. This option is very useful to compensate for minor imperfections in the metal sheet position and rotation.

- **Object Spacing**: This is the minimum distance to be set between nested parts.

Though the **Arrange** command is somewhat useful, there is a much better way to create nested parts inside Fusion 360.

The Nesting and Fabrication extension

The default version of Fusion 360 is a bit limited when it comes to nesting, but there is quite a good extension focused on nesting and fabrication that we can purchase to expand the nesting capabilities.

This extension, called **Nesting and Fabrication**, is intended to be professionally oriented, therefore it doesn't come with a hobbyist license. However, if you are interested, you can start a seven-day trial using the following link: `https://www.autodesk.com/products/fusion-360/nesting-fabrication-extension?term=1-YEAR&tab=subscription`.

> **Note**
> Such an extension is not mandatory for laser cutting; it is simply aimed at increasing nesting capabilities. If you don't plan to learn this extension, you can jump directly to the next chapter.

Now that the extension has been enabled, we can find all the commands related to nesting inside the **FABRICATION** tab:

Figure 12.10: The Fabrication tab

If your panel doesn't look like mine, chances are high that the plugin was not properly installed or licensed. If this is the case, you should restart Fusion 360.

Now it is time to unleash the whole power of the extension – in the following pages, we will explore most of the added commands to create advanced nesting management.

Material and sheet format

The first step is to specify the sheet format we want to use for our production. Therefore, from the **Manage** panel, launch the **Process Material Library** command:

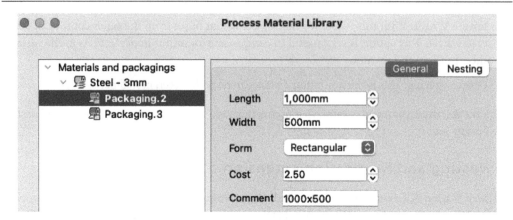

Figure 12.11: The Process Material Library panel

This window is the library where we can save and retrieve any metal sheet format. When first launched, we may have an empty library, but we can add more materials and formats for our daily operations. As we can see from the screenshot, at the moment, there is only a single material available, which is a **Steel** sheet, **3 mm** thick.

In addition to the material type, we can find elements called **Packaging** – those are basically the sheet format we can use. Since we decided to use a format of 1000x500 millimeters for the nesting, we can simply select one of the packages and then edit the **Length** and **Width** values on the right. This will ensure that all the nested parts will be placed inside the boundaries of our sheet.

In addition to the outer contour of our metal sheet, we can also specify an additional offset from the outer contour to prevent parts from being placed too close to the border. These options can be found inside the **Nesting** tab (which is to the right of the **General** tab):

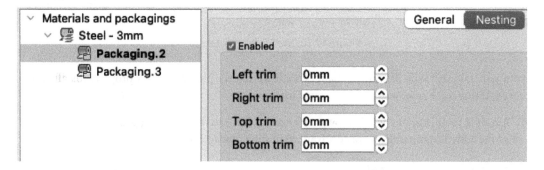

Figure 12.12: The Nesting tab (Packaging.2)

For our example, I won't add any offset, but such an option is especially useful when the metal sheet is not really squared or if there is a high degree of uncertainty about the placement of the sheet on the working bed.

Now that we understand how to set up the sheet format, we can move to the material itself. Let's click on **Steel - 3mm** to work with material parameters instead of packaging parameters:

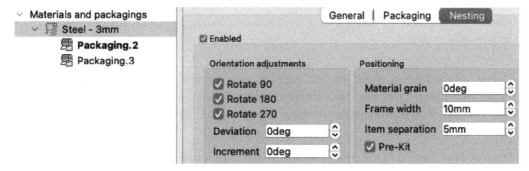

Figure 12.13: The Nesting tab (Steel - 3mm)

In this **Nesting** tab, we can change how parts can be rotated onto the sheet to pack them together for nesting. By default, Fusion 360 tries to arrange components inside the sheet boundaries by rotating them at 90° intervals; most of the time, this is more than enough for a good result. However, if our components have very elongated shapes or have many edges and angles with weird values, we may need to try smaller angle steps.

To reduce the angle step, set the **Deviation** angle to **90deg** and the **Increment** angle to a value as small as we see fit; for example, we may set it to **5deg** or similar.

> **Note**
>
> By default, Fusion 360 uses main rotation steps for part placement. As we can see in the previous screenshot, there are three options enabled: **Rotate 90**, **Rotate 180**, and **Rotate 270**, which are the main fixed rotation steps.
>
> **Deviation** will add a tolerance to the fixed rotation step. So, a **Deviation** angle of **90deg** will allow angles between 0° and 90° on each of the main rotation steps; this means that if all the main rotation steps are enabled, the part can be rotated from 0° to 360°.
>
> **Increment** is the rotation step used to find the best placement for the nesting.

Nesting is more or less like playing Tetris, having the option to rotate the bricks with small angle steps. As you may expect, using small angles to generate the nesting will take more time, but in some situations, it's definitely worth it. In our scenario, we don't really need small rotational steps since all the parts are squared; therefore, we can leave the default values (**0deg**).

Another option related to angles is **Material grain**; this option has been created to take into account a possible material **anisotropy**. Using this option, we can specify the direction of grains on the metal sheet; however, most of the time, we just have to leave the default value of **0deg**.

> **Note**
>
> A material is defined as isotropic if its mechanical properties along all three of its directions are constant. In contrast, anisotropy happens when the mechanical properties change between directions. Metal sheets tend to have anisotropies along the direction of the grains (lamination direction).

Moving on, by using **Frame width**, we can specify the minimum distance of the nested part from the border, while **Item separation** lets us set the minimum distance between parts. We will leave the default values for all these parameters; however, I thought it was important to review their usage to better understand the process.

Now that we have checked all the options to manage the sheet format and its material, we can jump to the part list!

The part list

I want to start off by saying that in this section, we won't modify any parameter for the example we are working on, so if you wish to skip ahead, you can. However, I think it is quite important to get a complete understanding of how a nesting placement can be modified for each component.

There is a single panel called **Component Sources** for managing parts for the nesting:

Figure 12.14: Sources location

We have already processed the material and its dimensions; however, we still need to specify which parts have to be included in the nesting and their requirements. This command lets us access a list of all the components in our assembly and allows us to override the parameters for each part individually:

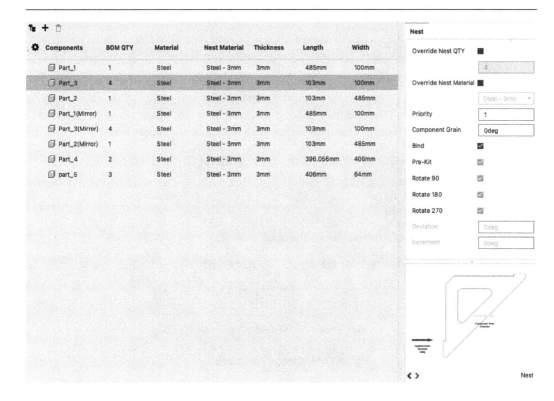

Components	BOM QTY	Material	Nest Material	Thickness	Length	Width
Part_1	1	Steel	Steel – 3mm	3mm	485mm	100mm
Part_3	4	Steel	Steel – 3mm	3mm	103mm	100mm
Part_2	1	Steel	Steel – 3mm	3mm	103mm	485mm
Part_1(Mirror)	1	Steel	Steel – 3mm	3mm	485mm	100mm
Part_3(Mirror)	4	Steel	Steel – 3mm	3mm	103mm	100mm
Part_2(Mirror)	1	Steel	Steel – 3mm	3mm	103mm	485mm
Part_4	2	Steel	Steel – 3mm	3mm	396.056mm	406mm
part_5	3	Steel	Steel – 3mm	3mm	406mm	64mm

Nest

- Override Nest QTY ■ | 4
- Override Nest Material ■ | Steel – 3mm
- Priority | 1
- Component Grain | 0deg
- Bind ✓
- Pre-Kit ✓
- Rotate 90 ✓
- Rotate 180 ✓
- Rotate 270 ✓
- Deviation | 0deg
- Increment | 0deg

Nest

Figure 12.15: The part list

Let's review the window step by step since it's quite rich with content:

- **BOM QTY**: This is the number of parts about to be used for the nest, and is useful when checking whether the calculated quantity matches our expectations. To change this value, we can use the **Override Nest QTY** option and change the number of parts to be cut. You may want to tweak this value if, for example, one of the parts previously cut featured defects or was lost.

- **Nest Material**: This is the material that the parts are going to be nested on. Please note that material, in this context, is intended as metal type and its thickness. In our example, all the parts have to be cut on a steel sheet 3 mm thick; that's our material then. As always, we can change this value as well using **Override Nest Material**.

- **Priority**: This sets the order of importance regarding which part to cut first. A high priority will put the part on the first sheets, while a low priority will put it on the last sheets to be cut. For our example, we don't need certain parts to be cut first; therefore, we can leave the default value of **1** for every component.

- **Add a new component** (+): This allows you to add external components to the nesting.

- **Component Grain**: This specifies a certain angle to orient a part with, taking advantage of the grain direction. By the way, if we want to check that the part is going to be oriented properly along the grain direction, we can check the placement preview on the bottom right of the window.

- **Bind**: This option synchronizes all the remaining settings with those specified inside the material setup we mentioned in the previous section. If we untick it, we can manually adjust those for each part, one by one. For the example study, we can leave this option checked.

We have now discovered all we need to know about setting up the nesting, and it is now time to move on to create one.

Creating the part nesting

The command to create our nesting is right on the main toolbar:

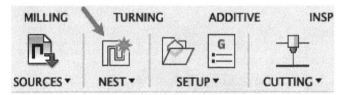

Figure 12.16: Nesting command

Let's click on the **Create Nest Study** command and explore its options. The first tab we have to configure is called **Study**.

The Study tab

In the **Study** tab, we can manage the number of components we want to create the nesting for:

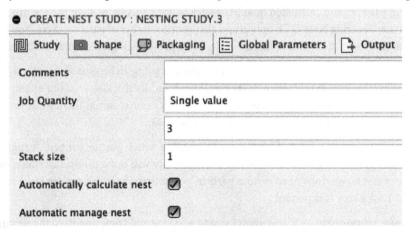

Figure 12.17: The Nesting Study tab

Let's suppose we want to cut enough components to be able to build three welded structures. To specify the total number of part kits we want to cut, we simply have to specify the value inside **Job Quantity**; in our case, we go for 3 then.

Using **Stack size**, we can set how many metal sheets we want to stack for our nesting; with laser cutting, we should always just use a single sheet; therefore, we must keep this value at **1**.

> **Note**
>
> **Sheet stacking** is a manufacturing technique sometimes used with waterjet machines. In most situations, the waterjet doesn't cut even nearly as fast as a laser machine can; however, to increase production, it is possible to stack multiple raw material sheets one on top of the other, and cut them at the same time.
>
> I suppose this technique sounds amazing, but can you guess why it is never done on a laser machine? Laser machines heat the parts up to melting point; therefore, having multiple sheets piled up will always result in welding them together!

The Shape tab

Inside the **Shape** tab, we can find another list of the components about to be placed inside the nesting:

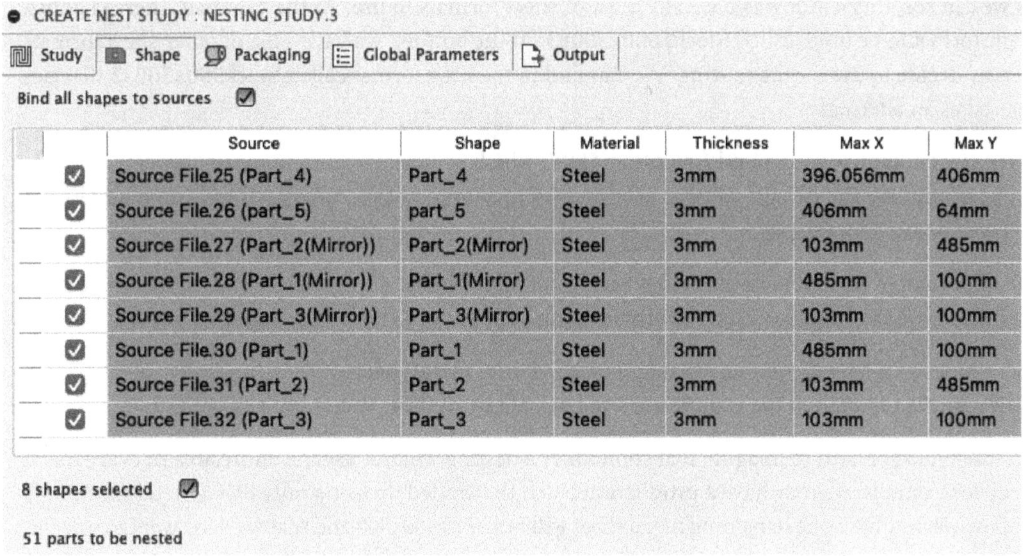

● CREATE NEST STUDY : NESTING STUDY.3

[ ] Study   [ ] Shape   [ ] Packaging   [ ] Global Parameters   [ ] Output

Bind all shapes to sources   ☑

		Source	Shape	Material	Thickness	Max X	Max Y
	☑	Source File.25 (Part_4)	Part_4	Steel	3mm	396.056mm	406mm
	☑	Source File.26 (part_5)	part_5	Steel	3mm	406mm	64mm
	☑	Source File.27 (Part_2(Mirror))	Part_2(Mirror)	Steel	3mm	103mm	485mm
	☑	Source File.28 (Part_1(Mirror))	Part_1(Mirror)	Steel	3mm	485mm	100mm
	☑	Source File.29 (Part_3(Mirror))	Part_3(Mirror)	Steel	3mm	103mm	100mm
	☑	Source File.30 (Part_1)	Part_1	Steel	3mm	485mm	100mm
	☑	Source File.31 (Part_2)	Part_2	Steel	3mm	103mm	485mm
	☑	Source File.32 (Part_3)	Part_3	Steel	3mm	103mm	100mm

8 shapes selected   ☑

51 parts to be nested

Figure 12.18: The Nesting Shape tab

If we properly checked the **Component Sources** panel, we shouldn't have any surprises here, so we won't look at this too much. The only option we will review is **Bind all shapes to sources** – this is quite a self-explanatory command, letting you auto-update the nesting in case any attribute of the material or its format changes.

The Packaging tab

Inside the **Component Sources** panel, we already specified the material and material thickness to use for our parts. However, at the moment, we haven't specified the material format (packaging) yet. This **Packaging** tab lets us do just that:

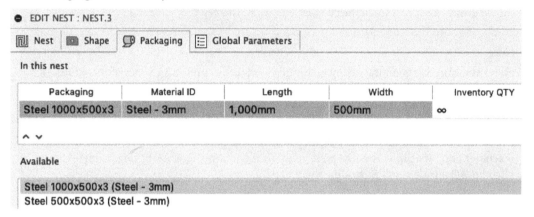

Packaging	Material ID	Length	Width	Inventory QTY
Steel 1000x500x3	Steel - 3mm	1,000mm	500mm	∞

Figure 12.19: The Nesting Packaging tab

As we can see, this window is basically a list of sheet formats to use. At the moment, there is only a single format to be used, called **Steel 1000x500x3**. At the bottom, we have a list of other sheet formats we may decide to use for the nesting. We can find in the list a format called **Steel 500x500x3** that can be used as an alternative.

In the right column, we can find **Inventory QTY**, which is the maximum number of sheets available in our inventory. The default value is infinite (∞), meaning that we have no limit to the number of sheets to be used for nesting.

The important thing to remember is that we are not forced to use a single sheet format for our nesting, but instead, we can combine multiple formats for the calculation. This is very important, so let's see why.

Let's suppose we have a company that, as standard, uses metal sheets with a dimension of 1000x500. If they are standard, then the warehouse will have plenty of these sheets.

However, it is not hard to imagine that sometimes a nesting will not use the entire area of every metal sheet; for example, we may have a production batch that ended up using only half a metal sheet. Let's suppose we now have one remaining metal sheet with a size of 500x500 and that we don't want to scrap it!

If only we could mix the sheet formats for nesting calculation.

Well, it turns out that we can. We can simply add a new packaging format (which is 500x500) to the list with **Inventory QTY** set to **1** and move it to the top of the list:

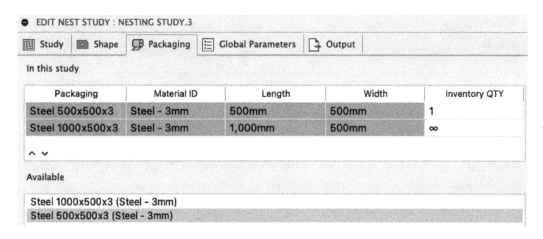

Figure 12.20: The Nesting Packaging tab

This way, the 500x500 sheet will be used first. Then, Fusion 360 will resume part placements on the standard 1000x500 format. This can save tons of money and reduce material waste!

Global Parameters

The **Global Parameters** panel is all about optimizing the parameters for placement performance:

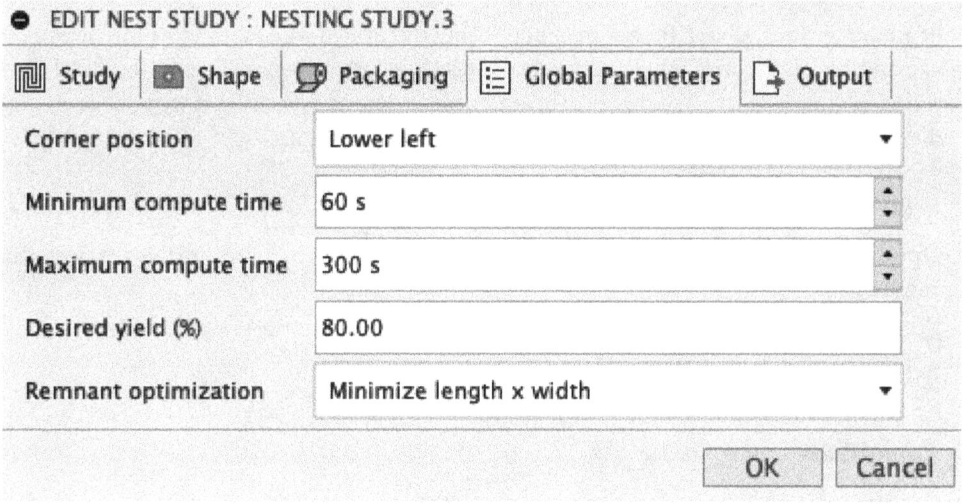

Figure 12.21: The Nesting Global Parameters tab

We won't change any of the default settings here, but we'll still review them for better awareness of the process:

- **Corner position**: This sets where we want to place the origin of each nesting.

- **Minimum/Maximum compute time**: This lets us set limits on the computing time for the nesting. The reason is quite simple: nesting is an intensive task for our CPU; it can take several minutes depending on how many parts are to be nested and their shape complexity. Fusion 360 will perform multiple attempts until the perfect optimization is reached. Sometimes, however, we may prefer to reduce computing time for a nesting pattern to be good enough. Most of the time (with basic components) we can leave these values unchanged.

- **Desired yield (%)**: This is the target nested area efficiency. So, what is nesting efficiency? It is a score that evaluates how much of the metal sheet is covered by parts. An efficiency of 100% means that the entire area is used, while an efficiency of 50% means that only half of the metal sheet is used. The default value is 80%, so if every sheet's total area is covered by at least 80% with parts, the nesting engine stops further optimization. We will use this default value.

> **Note**
> Inserting crazy high values (such as 99%) will result in a maximum compute time interruption since it isn't likely that any nesting could ever be so well optimized.

- **Remnant optimization**: If the sheet cannot be entirely covered with parts, using this option can manage the leftover area. This option lets us specify whether we prefer a leftover metal sheet with a minimum length, width, or both. For example, we may specify that we want a remnant sheet with measurements of 500x500 that we can use in the future.

The Output panel

The **Output** panel is the last panel we need to take a look at, and it is responsible for, well, the output results:

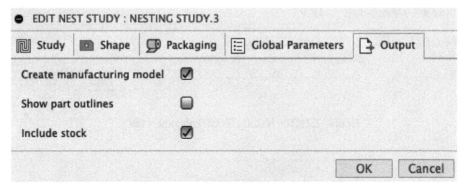

Figure 12.22: The Nesting Output tab

There aren't many options to set here, as we already handled most of the hard stuff:

- **Create manufacturing model**: This option should be ticked if we intend to create a proper toolpath inside Fusion 360; as we will do this in the next chapter, we will tick it. If we simply want to create a `.dxf` file with the contours of the nested parts, we can untick this option.

> **Note**
>
> Inside Fusion 360, it is possible to clone the 3D model developed inside the **DESIGN** environment and change it a little for manufacturing needs without breaking the link with the original. Such a clone is called a **manufacturing model**. In the case of laser cutting, we need a manufacturing model to arrange all the parts on a planar surface for nesting.

- **Show part outlines**: This is used to generate a `.dxf` file of all the nested parts. You should choose this if you want to use an external laser CAM other than Fusion 360. In our case, we won't use an external CAM; therefore, we can leave this option unticked.

- **Include stock**: This displays the sheet contours around the nested parts; it is really useful to get an idea of the overall placement; therefore, I always set it.

Now we have gone through all of those panels, it's now time to see the results:

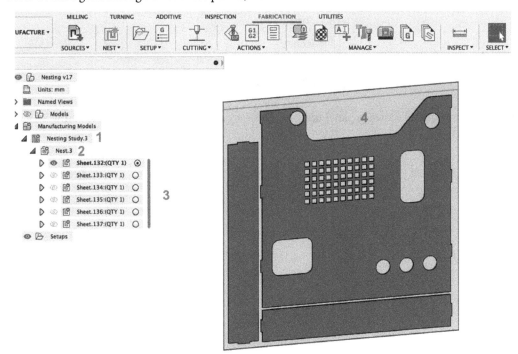

Figure 12.23: Results

As we can see, we now have a new manufacturing model (*1*) called **Nesting Study.3**, which contains a nesting called **Nest.3** (*2*). Inside the nesting, we can find six sheet layouts (*3*). Clicking on one layout will show how the parts are positioned on the sheet area (*4*).

We can also find more details regarding nesting. We just have to right-click on **Nesting Study.3** (note that your name may change) and click on **Compare**:

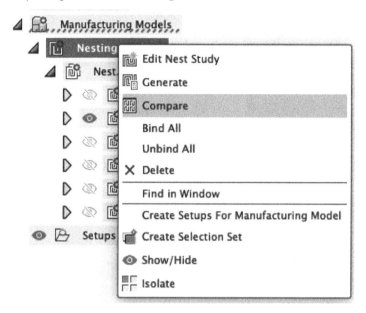

Figure 12.24: The Compare option

This will lead us to the following panel that contains everything we will ever need to know about the nesting:

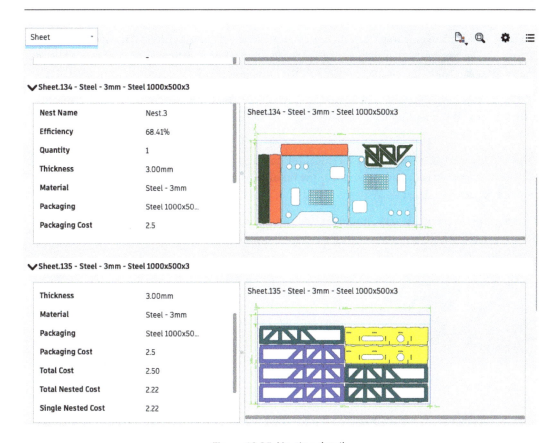

Figure 12.25: Nesting details

Here, there is a list of every sheet created for the nesting with its area **Efficiency**, its **Quantity**, and its **Packaging**. We can also find quite a useful preview with all the parts highlighted in different colors.

I'm pretty sure that now you have a better idea of why the Nesting extension is better than the Arrange command: it gives us much more power and flexibility!

Summary

That's the end of the chapter. We have now covered a large number of concepts and options for nesting. Let's briefly recap what we went through.

First, we analyzed our welded example part and the components it's made of. Then we found out how sheet formats and batch lot numbers can affect the efficiency of nesting.

After that, we found an easy way to create nested geometries using Fusion 360's **Arrange** command, though it's quite a simple and limited way to approach nesting optimization. So instead, we explored a Fusion 360 extension focused on nesting creation and covered all the options for material and packaging setup.

Lastly, we created the nesting itself, discovering how to reuse metal sheets only partially used by previous operations.

All the points and hints discussed are critical for competitiveness in the manufacturing market. I hope that, even if we only had a glimpse of the tip of the iceberg, this still lets you become a better and more prepared designer.

Preparing a good nesting is very important, but we still need to learn how to cut the nested parts. Therefore, in the next chapter, we will find out how to create a cutting toolpath for our laser machine using Fusion 360.

13

Creating Our First Laser Cutting Operation

In this chapter, we will find out how to implement a laser cutting setup and a cutting operation to process one of the nested layouts created in *Chapter 12*.

The goal of this chapter is to give you a general understanding of the process of laser cutting, highlighting its complexities. In the process, we will also discover how and why cutting tabs should always be added and how to compensate for the cutting width of the laser beam.

In this chapter, we will cover the following topics:

- Using Fusion 360 for laser cutting
- Creating a new setup for laser cutting
- Creating a new cutting tool
- Implementing our first cutting operation

Technical requirements

To follow along with this chapter, make sure that you have read *Chapter 11* and *Chapter 12*, as we are going to resume where we left off.

Using Fusion 360 for laser cutting

This chapter is intended to present all the features and commands that Fusion 360 offers. The goal is to create a cutting toolpath ready to be exported to our cutting machine. I aim for my explanations to be as clear and complete as possible; however, there are a few important limitations related to the subject.

As silly as it may sound, laser cutting is not similar to milling. In milling, there are multiple parameters ruled by simple formulas shared as standard by every milling machine brand. The chances are high that if we take a set of cutting parameters used to mill a part on a first CNC and use them on a different brand machine, they will work just as well. However, laser cutting is totally different. Every machine brand is different from its competitors and the number of variables to take into account is huge.

For a proper laser cutting operation, to list just a few parameters, we need to consider the material type, its thickness, its surface finish, and its behavior when exposed to a certain wavelength. In addition, we have to set the cutting gas to be used and the gas pressure, the shape of the nozzle, the laser power, and its pulses, not to mention the focal length of the lens and the focus position.

As you can imagine, all this stuff is really hard to measure or quantify and, for sure, is far beyond the skills of a mechanical engineer or a CNC operator (for sure, far beyond my skills!). That's why, to keep things simple, every brand provides an integrated solution for CAM with a complete database that covers every possible scenario.

Fusion 360 doesn't feature a cutting database; therefore, using it to generate laser cutting operations isn't a great idea; it is doable but quite difficult at first since we have to renounce cutting databases integrated with other programs.

Therefore, a much better option is to use Fusion 360 from part design up to nesting operations and then export the nesting layouts as .dxf files to other CAM programs centered on laser cutting.

Having said that, for the sake of completeness, during this chapter we will still use Fusion 360 to generate a toolpath, but we won't be able to optimize cutting parameters much. We will work with very basic settings that are good enough for a hobbyist but not much more.

Now that we have warmed up a little bit, we can begin with the laser cutting setup.

Creating a new setup for laser cutting

At the end of the last chapter, we created a set of nested parts; now, it is time to move on and create a new setup, a fundamental step before any cutting operation.

To create a new setup, we can launch the **SETUP** command from the **FABRICATION** tab:

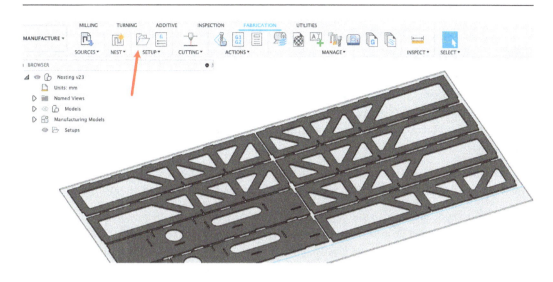

Figure 13.1: The Setup command

Once launched, we can explore the different tabs of the panel that appears.

The Setup tab

Inside the **Setup** tab, we can find all the options to select the components to be included in the cutting setup and the position of the coordinate system:

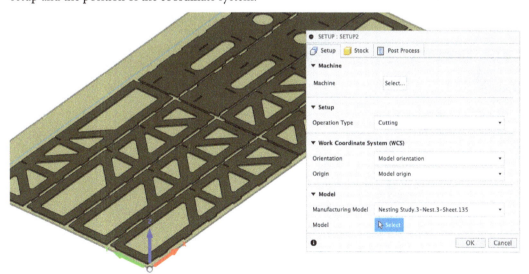

Figure 13.2: Setup's Setup tab

The first option to check is **Operation Type** – up until now, we have worked with **Turning** and **Milling**; this time, we shall set it to **Cutting**. Please note that a cutting setup can be used for laser cutting, plasma cutting, or waterjet; there is no difference for Fusion 360.

In the **Work Coordinate System (WCS)** subpanel, in the **Orientation** and **Origin** drop-down menus, we can set the position of the coordinate system and its orientation. This process is identical to what we already found for turning or for milling; we just have to check that the axis orientation and origin match our machine.

In the example, we can see that the Z axis is pointing upwards and that the X axis is pointing along the major dimension of the metal sheet, which is the typical axis configuration of most machines.

Inside the **Model** subpanel, we can find the options to pick the geometry to include in the setup. Using the **Manufacturing Model** drop-down menu, we can pick the nesting we plan to create the setup for. At the moment, the selected nesting is **Nest.3-Sheet.135**, which is one of the several metal sheets we created in *Chapter 12* (we can find the nesting layout on the left of the diagram).

Note

We have to create a setup for each metal sheet we plan to cut. Since these setups would be almost identical, in this chapter, we will create a single setup for a single sheet and forget about all the others.

Then, using the **Model** selection tool, we can individually select parts to be included in the setup. If we skip this selection, all the parts inside the nesting will be automatically included. In our case, we want to cut all the parts; therefore, we can leave the selection box blank.

The Stock tab

The **Stock** tab lets us specify the overall dimensions of the metal sheet we plan to cut:

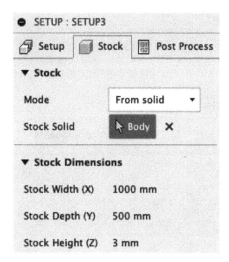

Figure 13.3: Setup's Stock tab

If you created the nesting using the Nesting and Fabrication extension, you don't have to change anything here since the stock will be automatically set correctly. Here, the **Stock Dimensions** measurements are already set to 1000x500x3, which is the packaging format we used for our nesting in fact.

Just in case you didn't use the nesting extension, you will need to manually specify the dimensions much as we did for the milling example.

The Post Process tab

Inside the **Post Process** tab, we can handle the **WCS offset** and give a name and a comment to the CAM program created by the setup:

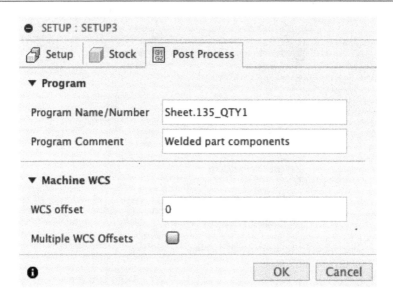

Figure 13.4: Setup's Post Process tab

A proper naming convention with multiple sheets is critical for remembering the sheet and the quantity to cut.

For **Program Name/Number**, I suggest you name it with a hint containing the sheet number and the quantity needed.

In our example, we need to cut six different sheets to build three welded parts; therefore, a proper naming convention for the G-code file may be something like `WeldedAssembly_Sheet1of5`.

That is all we need to know for the setup creation, so now we can press **OK** and move on to the next section, where we will create a laser tool to use for the cutting operation.

Creating a new cutting tool

Fusion 360 comes with quite a limited range of laser tools – actually, there is only *one* at our disposal. As you may guess, such a limited library spells trouble for us since, this time, there aren't plugins or libraries to use; the fastest way is to duplicate the existing tool and change it to our requirements.

To duplicate a tool inside the library, we can simply right-click on it, copy the tool using the **Copy tool** option, and then paste it inside the local library:

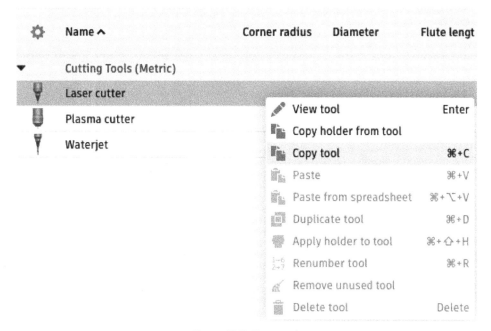

Figure 13.5: Copy tools

My suggestion is to rename the tool with a name that precisely describes the tool's intended use; in our example, I will rename it: **Fiber 4KW Steel 3mm O2**.

In order to edit the copied tool, we simply right-click on it and select **Edit tool**:

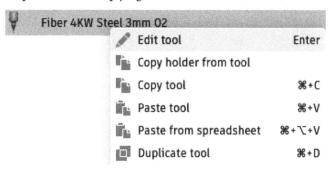

Figure 13.6: Edit tool

When editing the tool, we will focus on two tabs: the **Cutter** and **Cutting data** tabs.

The Cutter tab

First, let's take a look at the **Cutter** tab:

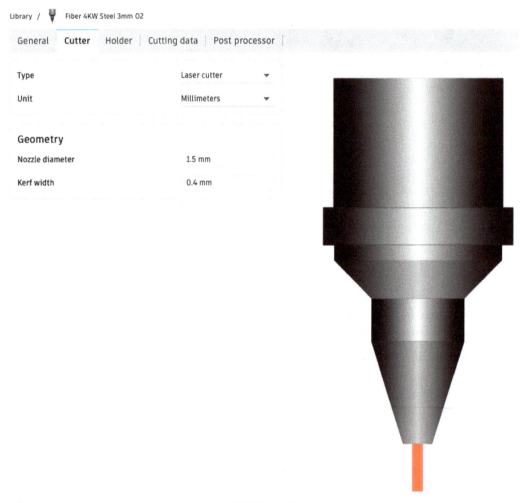

Figure 13.7: The Cutter tab

Inside the **Cutter** tab, there are two parameters we can specify:

- **Nozzle diameter**: This is the diameter of the hole where the laser beam and the jet of gas are emitted. A typical nozzle diameter can range between 1 mm and 4 mm depending on the type of cut. We will keep the default settings of **1.5 mm**.

- **Kerf width**: This is the cut width, which is the most important parameter to set since it drives the compensation of the cutting centerline. This parameter is highly dependent on the material thickness, the cutting power, and the lens our laser head is equipped with. However, considering a fiber source of 4 kW and a mild steel sheet of 3 mm, typical values may range between 0.2 and 0.4 mm, therefore we may keep the default value of **0.4 mm**. Since we don't have a database, this is a tentative value; after the first part is cut, we shall check the real width of the cut, measure the part's dimensions, and adjust this value accordingly.

The Cutting data tab

The **Cutting data** tab is quite important and yet difficult to set. I must confess that I'm not an expert on setting laser parameters without the support of a material database; however, let's go through this tab together. We will just review the meaning of most of the parameters; forgive me but I'm not skilled enough to suggest a proper value for all of them:

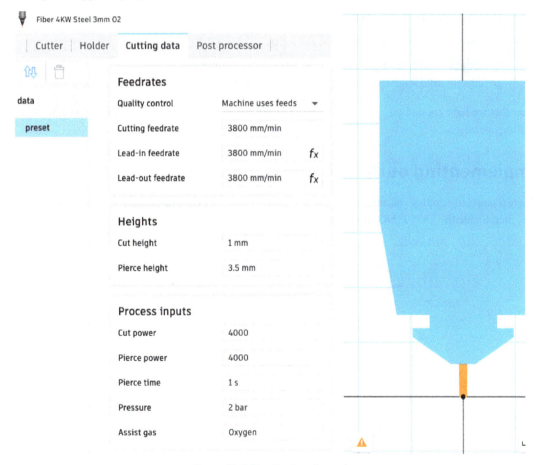

Figure 13.8: The Cutting data tab

Let's review all the points together:

- **Cutting feedrate**: This is the cutting speed that will be used in the operation. The default value is 1000 mm/min, which is definitely a very slow pace of movement. Considering that we are creating a tool to be used with a fiber laser of 4 kW cutting a 3mm thick steel panel using oxygen, I think that a feed of 3800-4200 mm/min makes much more sense. For the example, let's be conservative and go for **3800 mm/min**.

- **Cut height**: This is the distance between the metal sheet and the tip of the nozzle.

- **Pierce height**: **Piercing** is the first phase of a cutting process, where the laser beam has not yet gone through the entire metal thickness. So, this height option is the distance between the tip of the nozzle and the metal sheet when piercing.

- **Cut Power** and **Pierce power**: This is the power sent to the laser source during normal cutting and piercing.

- **Pierce time**: This is the amount of time the laser will require to go through the metal sheet.

- **Pressure**: This is the pressure of the assist gas.

- **Assist gas**: This is the type of gas to be used for the cutting operation. For our example, we are cutting mild steel; the fastest performances for this material are achieved using **Oxygen**.

Now that we have created our new laser tool, we can save and accept the options and get back to the cutting operation.

Implementing our first cutting operation

There is one single cutting command available, called **2D Profile**. We can find this in the **FABRICATION** tab, then inside the **CUTTING** drop-down menu:

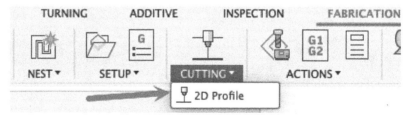

Figure 13.9: The 2D Profile command

Please note that in contrast to milling or turning, laser cutting is just a simple operation (there aren't complex machining strategies). Let's review the command together.

The Tool tab

As we have seen, the **Tool** tab lets us choose the cutting tool and its cutting speed:

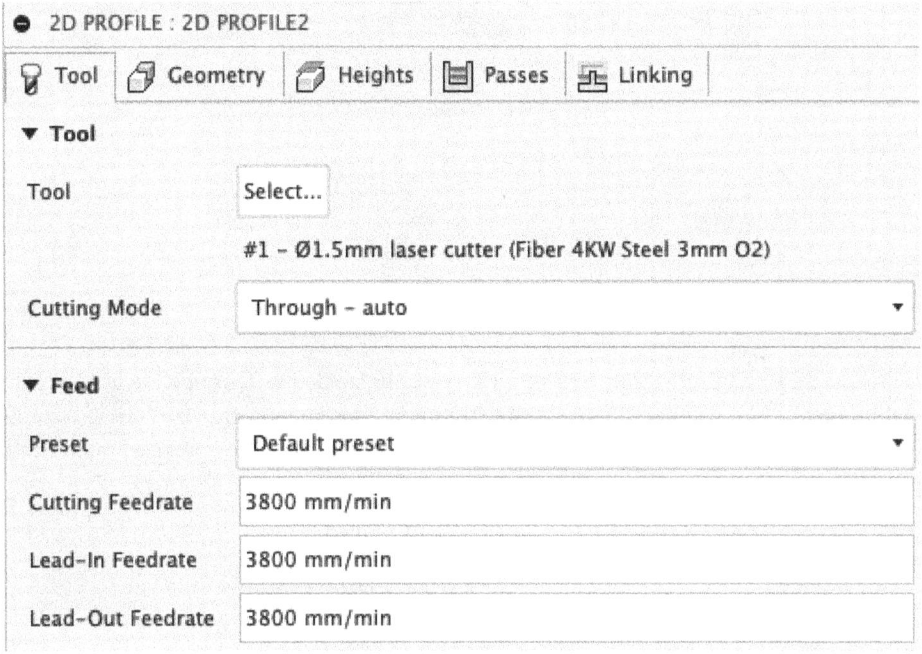

Figure 13.10: The 2D Profile Tool tab

First of all, we have to choose the cutting tool for the cutting operation. As we can see from the screenshot, using the **Select** option, we can pick the laser tool we just created (**Fiber 4KW Steel 3mm O2**).

Choosing the proper cutting tool is important, but we also have to set the **Cutting Feedrate** used for the operation. Most of the time, this requires us to manually insert the speed value.

However, since we already set the **Cutting Feedrate** during the tool creation, it is already retrieved inside this panel, and as we can find from the screenshot, it is properly set to **3800 mm/min**.

Now that both the cutting tool and the feedrate are set, we must be sure that the proper cutting mode is selected. The default **Cutting Mode** is called **Through – auto**; this is the default cutting preset to cut through the metal sheet (and therefore, it is the option we shall pick for the example).

However, let's take a look at the other options:

Figure 13.11: Cutting presets

As we can see, there are several cutting modes: all the options labeled as **Through** are used as cutting presets, while **Etch** lets us engrave the material without cutting it, and **Vaporize** is used to remove the plastic film on the top of certain sheets. All these different operations are managed by controlling the laser's intensity and its pulses, as well as the gas pressure.

The Geometry tab

This tab is one of the most important for the success of our cutting operation. Here we have to set the geometry to be cut and the tabs:

Figure 13.12: The 2D Profile Geometry tab

Let's review all the options available since they are quite critical:

- **Contour Selection**: This selection tool lets us pick the contours of the parts we want to cut.

- **Select Same Plane Faces**: When there are lots of components to select (which is most of the time), a much better option is to select a single contour and then flag this option to pick every other contour of the nesting. This is something we shall definitely do for this example.

- **Loops**: First of all, what is a loop? A loop is a closed profile inside the nested parts; it can be the outer contour that defines the shape of the part or an inner contour that defines the shape of a hole. This option lets us select which set of loops we want to cut with the cutting operation we are creating. We have three options to choose from:

  - **Inner loops**: Choosing this option will select only the closed holes inside our parts for cutting

  - **Outer loops**: Choosing this option will select only the outer contours of the parts for cutting

  - **All loops**: Choosing this option will select all the inner loops of a part and then its outer loop for cutting (this is the setting we should select right now)

> **Note**
> In cutting, the best procedure is to cut the inner loops of a part first and then its outer loop. Cutting the outer contour first will likely result in the part moving when cutting its internal geometry.

- **Side**: This option considers the kerf's width to adjust its position on the contour.

  As we just mentioned, any nested part is defined by its inner and outer loops. To be a little more specific, we can say these loops are a set of 2D lines that defines the theoretical boundaries of the part. Having said that, we have to remember that a laser beam, despite being quite small, has a non-zero width of cut. Therefore, in order to respect the dimensions of the part, we have to compensate for the beam position taking into account its size. As we can see from the diagram, the laser spot can be placed in the center, on the outside, or the inside of any contour:

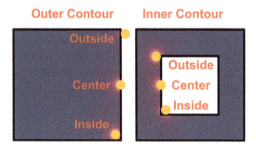

Figure 13.13: Centerline compensation

When cutting an outer loop, the laser will be moved on the outside of the contour; this way, the outer dimension won't be affected by the beam diameter. Vice versa, when cutting an inner loop, the laser should be placed on the inside of the loop; otherwise, the hole would be bigger than the theoretical value.

To consider both laser placement for outer loops and for inner loops, the best setting to choose is **Start outside**, which will work fine both for inner loops and outer loops.

The next set of options is used for tabs generation, but before taking a look at all the options, we will introduce what tabs are.

Tabs are programmed interruption of cutting with the scope of holding the cut part in place. When cutting with lasers, the metal sheet is not held in position with a clamping device like in milling; it is simply laid down onto the machine bed. Not only is it not held in place, but also the machine bed is made of sharp spikes that only touch the metal sheet on certain spots; therefore, cut parts are quite free to move and rotate:

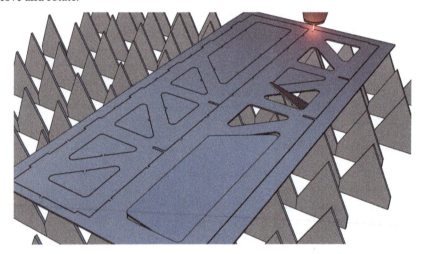

Figure 13.14: Tabs

As we can see in the diagram, there are two parts nested on the metal sheet. The part on the right is being cut without tabs; therefore, some of the inner loops already cut have moved or tilted, and they are not planar with the rest of the metal sheet anymore. This means that the laser nozzle may collide with them and get damaged, whereas the part on the left was cut with tabs and nothing moved, meaning there was no danger of collision.

Now that we understand the importance of tabs, we can review the options that control their creation and their geometries:

- **Tabs**: With this option, we can enable tabs to reduce the chances of colliding with an already cut part. Since we are cutting complex profiles, this is a setting we should always enable.

- **Tab Width**: This lets us choose how long tabs should be; longer tabs will hold better, but they will also require more effort to remove them manually. In our example, we will use **1 mm** tabs.

- **Tab Distance**: This is the distance between two tabs. The shorter the distance, the more tabs will be created. For our example, we will use a tab distance of **50 mm**.

The Heights tab

The **Heights** tab sets the position of the laser nozzle during the cutting process:

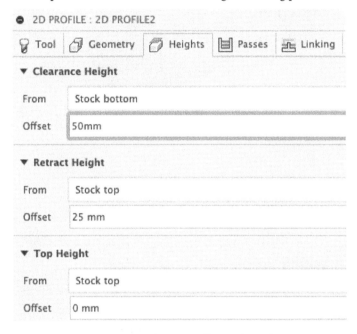

Figure 13.15: The 2D Profile Heights tab

The most important value to set here is **Retract Height**, which controls the position of the nozzle when not cutting. It is a safety measure to prevent the head from colliding with parts already cut; as we just discovered, laser cutting is a process where the parts tend to move and flip, causing collision dangers for the laser head.

Having used the tabs, however, we can be quite confident of having solved the problem, therefore a retract **Offset** of **25 mm** will be more than enough for our example part. All the other values can be left as default.

The **Passes** tab and the **Linking** tab can be left unchanged since the default settings are fine for 99.9% of cutting operations. Therefore, we can move to the next section and check the results.

Simulation results

After simulating the generated cutting toolpath, we should get something similar to the following screenshot:

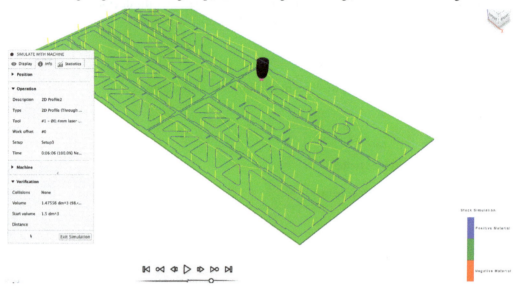

Figure 13.16: The results

As we can see, all the contours were cut, and we encountered no collisions.

In order to find potential issues, we have to look closer at the cutting path; in particular; we will always control the loops. We can zoom in close to the kerf to find whether the part dimensions are respected or whether the laser spot was misplaced.

As we found out back in *Figure 13.13*, the laser spot is very tiny but it generates a width of cut; therefore, we should always check whether the generated toolpath is positioned on the right side of the geometry.

Long story short, if we are checking an outer contour, the toolpath should be on the outside of the contour, while when we check an inner loop, we should find the toolpath on the inner side of the hole.

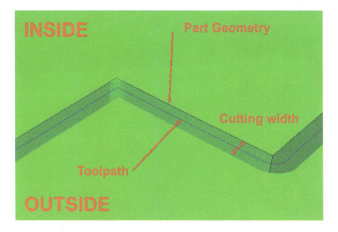

Figure 13.17: The cut details

Here, the result should be correct: the cutting toolpath has a constant offset on the outside from the part geometry (equal to half the cutting width). Therefore, the kerf is not reducing the part dimensions.

Now we can be quite confident that a proper toolpath has been generated. Congratulations!

Summary

This concludes the chapter and the section on laser-cutting technology. Overall, it has been quite a slim introduction to lasers since there aren't many options related to laser cutting to play with inside Fusion 360.

However, to recap what we went through here, we first found out why Fusion 360 is not the best solution on the market to implement cutting operations by highlighting most of its weaknesses.

Then we created our first cutting setup using one of the nested layouts we created during *Chapter 12*. While reviewing the options available for tool creation, we discussed a few important cutting parameters and their underlying importance.

Finally, we implemented the cutting operation itself and discovered the importance of tabs on the parts. Despite not being very detailed, I think this chapter was essential to get closer to the real world and its complexities.

It is now time to move to the next manufacturing technology: 3D printing!

Part 4 –
Using Fusion 360 for
Additive Manufacturing

In this part, we will focus on additive manufacturing, a technology capable of overcoming many of the limitations of conventional manufacturing processes. In doing so, we will present the different types of 3D printing machines, along with their pros and cons.

After an overall introduction to the topic, we will stick to the most common 3D printing process used by hobbyists and enthusiasts, FDM, and will try to find potential solutions to the typical limitations related to this process.

Once we have achieved a basic level of knowledge of 3D printers, we will return to Fusion 360 to create our first 3D printing operation.

Lastly, we will fully cover all the printing settings available inside Fusion 360 and learn their effects on the printing output.

This part includes the following chapters:

- *Chapter 14, Getting Started with Additive Manufacturing*
- *Chapter 15, Managing the Limitations of FDM Printers*
- *Chapter 16, Printing Our First Part*
- *Chapter 17, Understanding Advanced Printing Settings*

14

Getting Started with Additive Manufacturing

In this chapter, we will discover a new production technique, 3D printing, also known by its more technical name, **additive manufacturing**.

In contrast to every other manufacturing technology that produces objects by removing material, this manufacturing method creates a component by adding material – hence the name *additive* manufacturing.

This relatively new set of technologies has got cheaper and cheaper in recent years; as a result, 3D printing is now starting to overlap with traditional technologies for an increasingly wider range of production needs.

To become a better designer, it is important to be aware of alternatives to traditional legacy techniques. The goal of this chapter is to introduce you to the main printing technologies and their differences to help you understand which one may fit your needs best.

In this chapter, we will cover the following topics:

- Introducing additive manufacturing
- Exploring the pros and cons of 3D printing over conventional manufacturing processes
- Comparing different 3D printing technologies

> **Note**
> We will just focus on production scenarios related to the mechanical industry; however, please note that by the time this book reaches the shelves, new additive processes and solutions will have been released. Complete coverage of additive manufacturing is worth its own book.

Technical requirements

This chapter is a fresh restart on a new manufacturing technique; therefore, there aren't special requirements. However, having read *Chapter 1* and *Chapter 6* will enhance your understanding of additive manufacturing's potential.

Introducing additive manufacturing

Additive manufacturing is a relatively recent manufacturing technology, dating back to the 1980s. As always, it is a bit difficult to give strict definitions, but let's just say that most traditional production processes rely on removing unwanted material from a solid block of raw material (as a sculptor would do with a block of marble to create a statue).

Additive manufacturing is completely different; instead of removing material from a block, it focuses on creating the needed part by stacking very thin layers, one on top of the other.

Figure 14.1: An example of 3D printing

Since a 3D-printed object is created layer by layer, it is possible to create complex shapes not only on the outside of the part but also on the inside. Looking at the preceding figure, we can see an inner honeycomb pattern that reduces the overall weight of the part without compromising too much of its stiffness; such reinforcement can be created only with additive manufacturing.

This approach has so much potential, and it supports the creation of shapes otherwise impossible to manufacture.

Let's find out the advantages and drawbacks of 3D printing compared to a legacy manufacturing process. Then, later in the chapter, we will discuss the different printing technologies.

Exploring the pros and cons of 3D printing over conventional manufacturing processes

In this section, we will discuss most of the key features of 3D printing along with its advantages and drawbacks.

The pros of 3D printing

Additive manufacturing is not just a new manufacturing process that allows us to create parts in different ways. Additive manufacturing established a brand-new approach to part design that keeps opening new perspectives; it is a really complex field, rich with different technologies and applications under continuous development.

Complex shapes

One of the strongest arguments in favor of additive manufacturing is its flexibility to create very complex shapes, its potential extending way beyond the capabilities of any other technology. The wide range of feasible shapes achieved is so incredible that it has opened the doors to a brand-new design approach that is often paired with 3D printing.

Let me introduce you to **generative design**.

Giving a precise definition of generative design is not very simple, but what we can say is that it uses advanced software capable of optimizing the shape of a part. A typical goal of this approach is to minimize the weight of a part and maximize its strength, which is the type of extreme optimization sometimes needed in motorsports or aerospace industries.

But generative design shapes rely on very complex geometries that are impossible to achieve by conventional technologies, which is why it often comes hand in hand with additive manufacturing, as this is the sole process capable of creating such components.

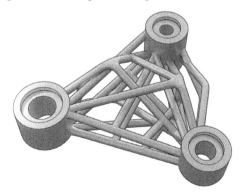

Figure 14.2: Generative design output

In the preceding figure, the shape of the triangle is very intricate and impossible for any conventional approach, but with 3D printing, it's not much more than a routine part!

To cut a long story short, whatever the shape, you can be confident that it can be printed (if you can afford it).

Rapid prototyping

Another benefit of additive manufacturing is its capability of delivering functional prototypes at an inexpensive cost compared to other technologies. Let's take the following injection-molded part as an example:

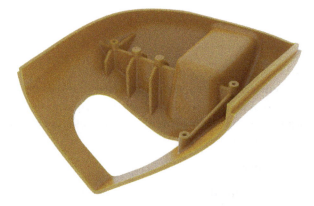

Figure 14.3: Injection-molding prototyping

The photo in the preceding figure shows quite a complex injection-molded part. As you may imagine, such a complex shape requires very complex (and expensive) molds.

Until not so long ago, the only way to create such a prototype would have been based on creating provisional molds for testing purposes. As you can imagine, prototyping is more about trial and error, so producing a mold for every prototype part can become quite expensive and unsustainable.

That's where additive manufacturing comes in – printing an object doesn't require any costly equipment, and it can produce multiple prototypes, each with its own shape. It should, therefore, come as no surprise that many people refer to additive manufacturing as **rapid prototyping**!

Production molds can easily cost several tens of thousands of euros (or even more, depending on complexities and dimensions), and provisional molds are always necessary to ensure that a final prototype can work flawlessly.

However, thanks to 3D printing, it is possible to create a provisional mold once the shape has been already optimized and passed through multiple 3D-printed steps.

This way, we can drastically reduce start-up costs by reducing the number of molds and equipment needed during the prototyping process.

Just in time supplying

What does **just in time** mean? It's a rather common approach of getting components without managing a large warehouse stock – components arrive at the company when they are needed or just before.

However, building or servicing complex assemblies is not a simple task; multiple elements have to be assembled together at the right time to sustain the scheduled production rate.

Let's suppose that we are a car manufacturer and that we have to assemble 100 cars every month. It turns out that our warehouse is out of stock on a certain component that must be installed in our cars in the middle of the assembly process. This is a very common situation nowadays for a variety of reasons: suppliers are late with their deliveries, a bad batch of components was rejected by the quality control office, and so on.

We definitely cannot afford to stop production waiting for components to arrive, as a company would lose lots of money. However, if such a missing part can be created via additive manufacturing, we may have the option to print it instead!

These days, there are several aerospace companies experimenting with 3D printing as a possible supply chain; instead of shipping a physical component overseas, it can be possible to just send a file to be printed. As you can imagine, such an approach has the potential of revolutionizing the industry once and for all; however, it's not a well-established supply chain yet.

Composites

A **composite** is a material where two or more different materials are combined to achieve better mechanical performances than the originals.

A typical example would be carbon fiber objects – they are made with several layers of carbon fiber cloth, Nomex honeycomb, and resin that hold everything together. These objects are very light and strong, which is why they are used in Formula One cars; however, they are also incredibly expensive and difficult to laminate.

3D printers can create composite objects too – a printed composite material is based on simple polymers, such as ABS or Nylon, reinforced with short carbon fibers, which further enhances the mechanical properties of the component.

The overall strength of a 3D-printed part reinforced with carbon fiber won't be as good as a laminated part created with traditional techniques, but it's still way better than plastic only!

In the following figure, there are two components; the one on the left has been 3D-printed using carbon reinforcements, while the one on the right has been laminated with a traditional process using carbon fiber sheets.

Figure 14.4: A carbon fiber duct

I really love carbon, and I risk digressing too much, so let's just say that a traditional carbon fiber component usually has better properties than a 3D-printed counterpart, but it is also way more expensive and time-consuming, since most of the lamination process relies on handmade steps. On the other hand, a printed carbon part can be the sweet spot between strength, weight, and costs.

Therefore, having the option to print using composites is the cherry on the icing on the cake that you don't want to miss.

The cons of 3D printing

Today, it is still unlikely to find a 3D-printed part in our daily lives, as additive manufacturing is not a mass production technique. Unfortunately, it probably never will – let's look at the main limitations that stop this technology from becoming more widely used.

Scalability

Scalability is the biggest issue related to additive manufacturing; not only is printing a part this way not cheap but also the cost is almost constant, whatever the production lot volume. Printing a single part, tens of parts, or tens of thousands of parts will always cost almost the same; it is not possible to reduce the cost significantly.

The cost of a printed part can be composed of the following components:

- **The print material**: The cost of the material composing the printed part.
- **The support material**: The cost of the additional geometries used to stabilize the printed parts. These structures are needed for certain additive processes and complex geometries.
- **The working time of the machine**: The cost per hour of the machine being used.
- **Electricity**: The cost of the power used to print a part.

As we found out, there is no startup cost or equipment cost to distribute over the number of parts produced. This is a strong point in favor of additive manufacturing with small production volumes, but it is also its Achilles heel with mass production.

Let's consider the following component being produced using both additive manufacturing and injection molding:

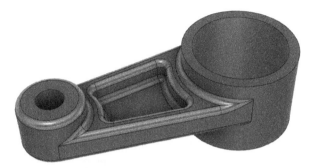

Figure 14.5: An example component

Both technologies are more than capable of realizing the shape of the part. The cost of the printed part can be influenced by a variety of factors, such as printing technology, material type, surface finish, and so on, but let's suppose that printing this part using additive manufacturing may cost around 10 euros.

On the other hand, injection molding has a huge cost, due to the initial mold creation.

However, such an important cost can be diluted on the parts – if the production numbers are high enough, the cost of the mold will increase its price tag by a very tiny fraction.

Imagine that we have to produce a component whose mold costs around 10,000 euros. Let's find out what happens when diluting the mold cost into the production volume:

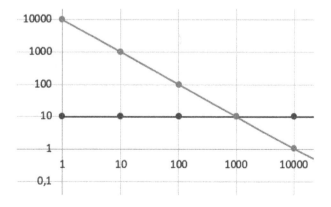

Figure 14.6: The molding cost versus production volume

On the *X* axis of the chart, we can find the production volume, while on the *Y* axis, there is the cost of a single component. There are two lines – the orange sloping one depicts the production cost for molded parts, while the blue horizontal one is additive manufacturing.

On the left of the chart, we can see that injection molding is way more expensive than additive manufacturing, since the mold cost is heavy when distributed on just a few parts. However, once the cost of the mold is split over a large number, we reach a break-even point with additive manufacturing (in the example, around 1,000 parts). For larger amounts of components, injection molding will be cheaper, and cheaper than 3D printing.

That's why additive manufacturing is never going to be a mass production technology – most of the time, it is simply more expensive than other technologies.

A limited range of materials

Another disadvantage of additive manufacturing is the range of materials available for printing. Don't get me wrong – today's 3D printers are printing everything, from metals to human cells, and so there is quite a large library, but it is still limited.

Limited part dimensions

Additive manufacturing is a rather complex process that requires a finely controlled environment; that's the main reason why 3D printers, generally speaking, do not have large build volumes. As you can imagine, there are exceptions – someone out there is probably printing jet turbines or even houses; however, generally speaking, we can be sure that a part with boundaries of 500 x 500 x 500 mm is going to be quite difficult to print in one piece only.

This is due to the fact that most printers on the market have a build volume of less than 250 x 250 x 250 mm (and sometimes way less than that). This volume limitation is quite restrictive for several mechanical components. There is always the option of slicing the part into multiple sub-components to glue together, but the process is never flawless; that's why I decided to include build dimension as a drawback for additive manufacturing.

This is all that you need to know about additive manufacturing. To cut a long story short, additive manufacturing can achieve terrific shapes at compelling costs for small batches; however, it is inferior to other production processes if higher quantities are needed or the part dimensions are quite large.

Let's now jump to the next section, where we are about to discover the main differences between the three most common additive processes.

Comparing different 3D-printing technologies

Choosing the proper additive technology for our project is essential if we want it to seriously compete with other production techniques. That's why we are about to discuss all of them in detail now.

Fused deposition modeling

Fused Deposition Modeling (FDM) is the cheapest additive manufacturing process and, therefore, is the most common technology used by small companies and hobbyists.

In the following diagram, we can find a basic simplification of the process:

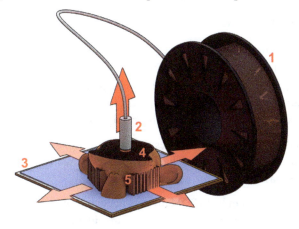

Figure 14.7: The FDM process

First, the printing material is supplied via a large spool of plastic filament (**1**). This filament is then forced into a hot end extruder (**2**) that melts the plastic and deposits a thin layer of material on the build plate (**3**).

Note that the interior of the part (**4**) is not solid, so by using this technology, it is possible to create a honeycomb infill. This type of infill pattern reduces the overall weight of the part (and material usage) without compromising its strength. (Check out *Figure 14.1* for a closer look at the infill pattern.)

Also, note that we are also using support material – in the diagram in *Figure 14.7*, we can find multiple pillars of printed material (**5**), with the sole responsibility of supporting yet-to-be-printed overhang areas.

The last thing I'd like to highlight, as silly as it sounds, is that the motion of the extruder relative to the build platform happens on the three main axes (*X*, *Y*, and *Z*). As we are about to discover, not every other additive process relies on this type of movement.

Now, let's quickly review the most important points to understand FDM technology by looking at its advantages and disadvantages. The advantages are as follows:

- **Multi-material support**: The material is fed via a spool and printed using an extruder; therefore, thanks to the low complexity of this process, it is possible to have multiple extruders working together. Using multiple nozzles, each extruding a different material in parallel, lets us create multi-material objects. For example, by using a stiff filament and a rubber filament, it is possible to print a complete toy car wheel with the tire and the rim in one shot! This is the only printing technology that supports multi-material objects.

- **Very cheap and simple**: An FDM 3D printer is a rather basic machine; it is based on a moving platform, a heating cartridge, and a nozzle. No lasers or complex sensors are required; as long as the plastic is melted at the right temperature and with the right flow, it will print just fine.

- **Large build volumes**: Thanks to their simplicity, FDM printers tend to have a rather large build volume compared to other 3D printing technologies.

The disadvantages of FDM are as follows:

- **Surface finish**: Despite having improved a lot in recent years, it is still possible to spot an FDM-printed part at first glance. This type of additive technology always leaves visible markings on the sides of the part, making it look a bit cheap compared to other additive processes. Using FDM denies a part any glossy finish or transparent look; therefore, this type of technology is often used for hidden structural parts or early prototypes only.

- **Support material**: FDM printing requires the positioning of support material to sustain overhanging areas of the part. These pillars have to be manually removed after the print is complete, and they basically become material waste.

- **Details and resolution**: Not only do FDM printers not grant a great surface finish, but they also aren't capable of printing very small details or reaching tight tolerances, which is something to consider when prototyping functional parts of a mechanism!

Now that we've had a look at this first additive technology, let's jump to the next one – stereolithography.

Introducing stereolithography

Stereolithography (**SLA**) is by far my favorite additive process. Let's find out why I love it so much and what it relies on by looking at the following diagram:

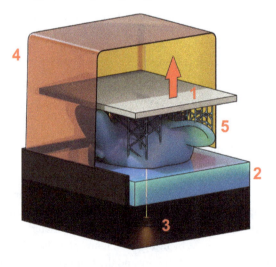

Figure 14.8: The SLA process

As we can immediately spot from the picture, an SLA 3D printer is rather different from an FDM machine. First of all, the build platform (**1**) just moves up and down and the part is printed upside down. The printing material is a special liquid resin sensitive to UV radiations (you can find out more about this in *Figure 11.1*), which is pumped inside a tank (**2**) where the part plunges while printing. On the bottom of the tank, there is a UV light source (**3**) focused on the bottom of the part; controlling the UV light, it is possible to solidify the geometry layer by layer.

Since the resins are sensitive to UV light, this type of printer features UV screen protection (**4**). This cover is not intended as a safety protection against our eyes but, rather, a protection for the liquid resin against sunlight! Since the resin is UV-sensitive, normal light can easily trigger **polymerization** and ruin the printing process.

> **Note**
> To keep things simple, let's just say that polymerization is a chemical process where molecules get bound together with a certain arrangement to create a polymer-like plastic.

Like the previous printing process, we can find support material structures with complex shapes (**5**) that support overhanging areas and that connect the printed part to the build plate.

Let's now find out the most important features of this technology, again by looking at the advantages and disadvantages. The advantages are as follows:

- **Surface finish**: This is, hands down, the best printing quality achievable by an additive manufacturing process. Stereolithography not only grants very smooth and shiny parts but is also the only 3D-printing technology that can print transparent components!

- **Highest precision**: SLA not only excels in terms of surface finish but also prints very tiny details with great precision, which is why it is often used for miniatures.

- **Materials**: In recent years, the number of available resins has increased significantly, and several third-party companies are offering multiple solutions to cover a wider range of mechanical, thermal, and aesthetic properties.

The disadvantages of SLA are as follows:

- **Costs**: Among all the additive manufacturing processes, stereolithography is probably the most expensive. Not only is this type of 3D printer definitely not budget-friendly, but also one kilogram of resin is way more expensive than a kilogram of any other additive manufacturing printing material.

- **Support material**: Similar to FDM, this process requires support structures that have to be manually removed before post-processing begins.

- **Curing**: Once finished, the printed part is covered in unpolymerized resin residue that must be washed with an alcoholic solution. This wash creates a special waste difficult to dismiss and is definitely not disposable with water down your sink! Depending on the resin, after this alcoholic bath, most of the time the resulting part remains sticky. Therefore, the part has to be cured with special UV lights that complete the polymerization. This is quite a long and complex post-processing of the parts – in my opinion, it is the worst drawback of this technology.

We just discovered that stereolithography is not as straightforward as other technologies and requires a long post-production process to achieve the final part. However, after all these steps, we can enjoy a very nice print result!

Now, let's look at our last additive technology – selective laser sintering.

Introducing selective laser sintering

Selective Laser Sintering (**SLS**) is probably the most flexible process and has several key differences from the others worth analyzing. In the following diagram, we can find a basic illustration of an SLS 3D printer:

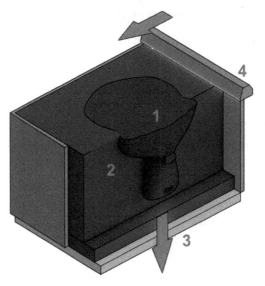

Figure 14.9: The SLS process

With SLS, the raw material used for printing is not resin or a plastic filament, it uses nylon powders instead. The printed part (**1**) is created inside the build volume (**2**), which is gradually filled with new layers of powder. Layer by layer, the build platform (**3**) moves toward the bottom of the machine, and a new layer is deposited on top of the previous one by a special distributor (**4**). A laser is then responsible for heating and melting the powder particles together, forming a solid part.

This is a really smart process not affected by many of the problems of other processes. Let's take a look at some of its advantages:

- **No support material**: Since the part is printed inside a volume of powder, it doesn't need to touch the build platform in any way; therefore, support structures are not needed. This is a huge advantage, since we don't have to worry about support material and how the support structures will connect to the build platform, meaning the entire build volume can be used to print as many parts as it is possible to fit inside, increasing productivity (the highest among the other 3D printing processes).

- **Most complex designs**: With this type of printing process, it is possible to create working mechanisms straight from the build volume; using SLS machines, we can print entire assemblies, and as long as there is a layer of powder between components, they won't jam. This is not as easy as it sounds, but with a simple mechanism, it is definitely doable.

- **Surface finish**: The surface finish, though not as good as most SLA prints, is generally good.

- **Material cost**: Since the print material is nylon powder, it is cheap, reusable, and can be recycled easily.

The disadvantages of SLR are as follows:

- **Cost**: Between the three technologies presented, an SLS printer is the most expensive

- **Printing material library**: The available choice of materials to print is quite limited, being mostly bound to nylon and a few other polymers

Now that we have explored all these interesting printing technologies, I'd really like to know which one is your favorite!

Summary

This concludes the chapter – I hope that your journey with additive manufacturing will be a source of new ideas and a starting point to discover different approaches to part design.

Let's recap together what we discovered during this chapter. At first, we discussed what additive manufacturing is and how it is different from other legacy production processes. After this, we explained in which scenarios additive manufacturing should be considered a possible alternative to other technologies. Most notably, we found out why 3D printers are a valuable solution for prototypes and the advantage of additive flexibility in just-in-time supply chains.

In the final section, we analyzed one by one the most common printing technologies (FDM, SLS, and SLA), comparing their features and drawbacks. Understanding all these points is very important; sometimes, a designer is bound to old ideas or old shapes driven by legacy manufacturing techniques. Grasping the innovation of additive manufacturing can lead to new solutions and new unforeseen scenarios.

Now, follow me to the next chapter, where we are about to discuss how to optimize the setup process of an FDM part.

15
Managing the Limitations of FDM Printers

We just discovered that additive manufacturing is quite an incredible technology, capable of creating any shape. But despite being very flexible, 3D printing is still bound to limitations that we must consider when approaching a new part study.

In this chapter, we are going to try answering the following question: can an FDM 3D printer create every imaginable geometry? The answer is... almost. We must understand that this type of printing technique has some limitations – some related to part shape, others to material properties.

In this chapter, we will cover the three main limitations of FDM printers that we have to consider:

- The first limitation is printing overhang areas; some geometries can be a bit challenging to print if not properly supported

- The second limitation is bed adhesion, which is related to the first layer placement; if the contact area between the printed part and the build platform is not large enough, the part might vibrate or detach from the platform while printing

- The third limitation, which is a bit more technical, is about anisotropies

After presenting these limitations, we will try to find a possible solution to these common issues by setting up a proper part placement or orientation.

However, as we are about to discover, sometimes it is not possible to fulfill every requirement, and then we will have to find a compromise. Understanding how to set up a component for printing can make the difference between a successful print and a waste of time and money. That's why this chapter should be read carefully!

In this chapter, we will cover the following topics:

- Printing overhang geometries
- Understanding bed adhesion
- Understanding anisotropies of the printed part
- Choosing the first layer placement and part orientation

Technical requirements

The only requirement for this chapter is to have read and understood *Chapter 14* – in particular, the section about FDM printers.

Printing overhang geometries

Overhang areas are sloped faces that protrude beyond the base of our model. We actually already faced a similar concept, undercuts, when studying milling issues:

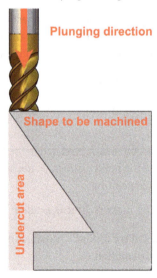

Figure 15.1: Undercuts for milling

As we may recall, undercuts are geometries that cannot be reached by the cutting tool, they are a huge limitation to milling processes even if we already found several possible solutions to fix undercuts in *Chapter 7*.

In this section, we will focus on additive manufacturing to find out whether undercuts are an issue that affects FDM printing as well, and if so, how.

Facing overhangs (undercuts) in 3D printing

Before jumping to the answer, let's get back to basics – additive manufacturing, instead of removing material from a solid block, creates a shape by stacking multiple layers one on top of the other.

For this reason, generally speaking, 3D printers don't suffer undercut-related issues nearly as much as milling or other legacy technologies. Let's investigate a little bit more into why.

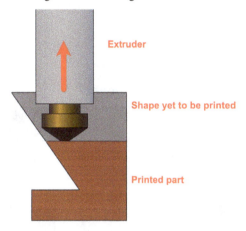

Figure 15.2: Undercuts for FDM

As we already know and can see in the preceding diagram, the part is created layer by layer. While the printing process is in progress, the extruder gradually moves upward, leaving lower geometries untouched. This is the reason that there aren't accessibility issues – the sloped face is not yet printed when the extruder moves in that area!

Unfortunately, there is a catch – under certain circumstances, undercuts can lead to poor quality in our prints. Let's check out a typical example:

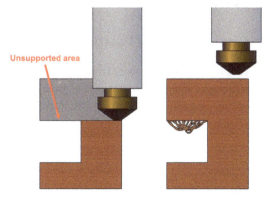

Figure 15.3: A sudden section change

This time, on the left, the shape to be printed has a sudden section change instead of featuring a sloped undercut face. This is a huge problem for an FDM printer because this type of additive technology relies on stacking a layer on top of the last one – it cannot print a layer on top of thin air!

For this reason, as we can see, the unsupported area will never be printed properly; the bottom face will feature extruded layers bent and distorted toward the bottom.

In the following diagram, we can see a wider range of overhang geometries and their resulting prints:

Figure 15.4: Different undercuts angles

The geometry on the left has a surface with a draft angle of around 45° – most of the time, this is considered the limit for a successful print. Steeper faces (such as the center geometry) or upside-down faces (the right geometry) cannot be printed properly.

So, now we understand the first FDM limitation – overhang areas are a bit challenging to print. But don't be frustrated, as this limitation doesn't have to restrict our imagination for new shapes. We still have an ace up our sleeve – support material.

Improving our prints using support structures

To cut a long story short, in order to avoid the extruder from extruding material over thin air, it is possible to create dispensable structures with the sole purpose of giving the extruder a layer to print onto. In the following figure, we can see good use of support structures:

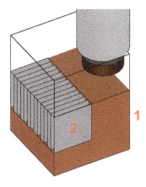

Figure 15.5: Support structures

Here, the part (**1**) is being printed, and the extruder is about to encounter a sudden section change, similar to what we analyzed in *Figure 15.4*. This time, however, instead of printing over thin air, there is a block of thin support structures (**2**) that will sustain the following layers.

These structures (generally referred to as support material) have to be manually removed once the print is complete. Therefore, they must be stiff enough to support the upper layers but also fragile and light enough to be removed without damaging the printed part.

> **Note**
>
> You may also have noticed from the diagram that it depicts two different colors – a darker one for the printed part and a lighter one for support material. With FDM printers featuring multiple extruders, it is possible to print support structures with water-soluble material – for example, PVA – that will leave no residue after being submerged in water.

As simple as it may sound, support structures are an optional help to support overhang geometries. If we can find a part placement that doesn't need support, then choose that part orientation instead.

If we print unnecessary support structures, there will be some downsides: the overall printing time will be longer, we will waste more material on the same object, and we will likely have a worse surface finish on the supported areas.

> **Note**
>
> Rule #1: Reduce support structures as much as possible.

As we will discover in *Chapter 16* and *Chapter 17*, creating simple support structures is not a hard task, and it can be performed automatically by the software.

Something that should be carefully considered, however, is setting and adjusting part placement. We will try to understand the basic concepts of part placement in the next section.

Understanding bed adhesion

Every FDM printing process relies on printing the part from the build platform of the 3D printer and stacking layers on top of it. As you can imagine, having the object perfectly glued to the build surface is super-important. It must not detach from the platform nor vibrate while printing; otherwise, the printing process will fail miserably.

Usually, the best prints are achieved when the area of contact between the build platform and the part is large enough to give perfect adhesion to the latter.

This is probably the most important rule for a successful print, and luckily, it is also quite simple to understand – we always have to put the larger and flattest area of our component in contact with the printing bed.

The reason behind this rule is quite intuitive; since the part, while printing, has to remain perfectly fixed to the printing bed, a larger contact area will grant better adhesion forces to counteract vibrations and local deformations. Let's find a practical example that will further clarify this golden rule for FDM printers:

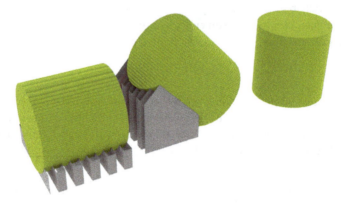

Figure 15.6: Different printing orientations

In the preceding figure, there are three different cylinders printed and oriented in different ways on the printing platform:

- The first cylinder is laid down on the rounded side of the cylinder, meaning that the contact area between the component to be printed and the platform is very tiny (it's a thin segment).

- The second cylinder is tilted and touches the build platform on an even smaller area; it's a single point of contact between the bottom circle and the platform. Such a tiny area is way too small and unstable to be used as the first layer of contact. On the off chance that we do get a successful print, it will likely feature a very bad surface finish and distorted layers, to say the least.

- The third cylinder is the only proper way of orienting this shape, done by laying it on the base of the cylinder; this will provide much better adhesion to the build platform and won't require support structures to have a chance of success.

> **Note**
> **Rule #2**: Large flat areas must be laid on the bed plate.

Now that we have covered this second golden rule, we can move on to the last one, which is a bit more technical and complex to understand, since it involves material properties.

Understanding anisotropies of the printed part

FDM prints feature anisotropies along the printing direction. Typically, the mechanical strength between stacked layers is weaker than the strength in other directions; therefore, the higher loads should always be applied along the layers, not perpendicular to them.

We should always consider these anisotropies when studying part placement, especially if we are about to print a functional prototype that will be loaded by forces and deformations. Let's take the following component as an example; this part is similar to a circlip ring and will act a bit like a spring:

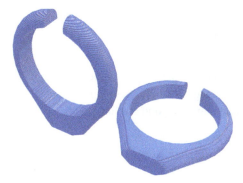

Figure 15.7: Another example of different printing orientations

As you can see, there are two different part placements for this part; the first one on the left features the part standing up, while the second shows the part laid on its side.

Both part placements are feasible from a printing perspective. However, since this is a component that will have to sustain forces, there is only a clear winner. Let's analyze the different layer orientations to see which is the toughest:

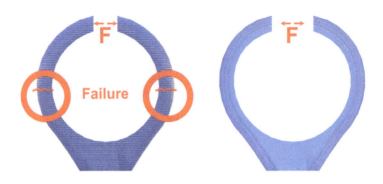

Figure 15.8: The points of failure

On the first layer orientation, the force (labeled **F**), applied as illustrated, will generate high stress on the two arms of the part, likely resulting in a static failure of one of the two, with a crack parallel to the layers.

On the other hand, the second layer orientation will result in a much more flexible component that will be able to sustain higher abuses!

> **Note**
> **Rule 3#**: Be mindful of the layer orientation.

As we just discovered, choosing the best printing placement is not an easy task – it largely depends on the part itself, and there is no magic formula that helps us decide. Therefore, in this chapter, we will try to give best practices that may be a good starting point.

Please note that sometimes it is not possible to respect all these suggestions – sometimes, one is in open conflict with another, so we will then have to find a compromise. In the following section, we will try to find the best printing approach for a rather simple shape, while trying to find the best placement that respects all the given rules.

Choosing the first layer placement and part orientation

After reading the previous suggestions, you may find yourself a bit disoriented with all of the different limitations. So, considering those, let's try to find the best part placement for the following component:

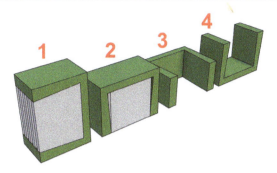

Figure 15.9: Choosing the best printing orientation

Here, we can find all the possible printing orientations for the given part. Let's analyze them and test whether the given rules are respected.

First, let's look at placement **1**:

- This placement requires quite a large volume of support material; therefore, the first rule is not really respected

- The first layer is quite large, and it will grant a good adhesion to the part; therefore, the second rule is respected.

- This type of layer placement will make the thin vertical wall very fragile; therefore, the third rule is not respected

Let's look at placement **2**:

- This placement requires quite a large volume of support material; therefore, the first rule is not really respected

- The contact area is very small; therefore, the second rule is also not respected

- Using this layer orientation, the horizontal wall will be flexible, but the vertical ones will be fragile; therefore, the third rule is not respected

Let's look at placement **3**:

- There is no support material; therefore, the first rule is respected

- There is quite a large area of contact between the part and the platform; therefore, the second rule is respected

- With this layer orientation, the thin walls are quite flexible; therefore, they won't break as easily as the other placements

Let's look at placement **4**:

- There is no support material; therefore, the first rule is respected

- There is quite a large area of contact between the part and the platform, so the second rule is respected

- Using this layer orientation, the horizontal wall will be flexible, but the vertical ones will be fragile; therefore, the third rule is not respected

Having analyzed all the possible placements, we can safely say that the best way to go is placement **3** – it can be printed very well thanks to a large contact area, won't require support, and will also feature good mechanical properties.

Note that real-life parts are more difficult to make than a simple squared part shown in a book on Fusion 360 CAM, but the basic rules are still the same. However, sometimes, we will have to find compromises or change the printing technology, with SLS being the most suitable for the most complex shapes.

Summary

That's the end of the chapter. Let's quickly recap what we explored.

First of all, we discovered overhangs in 3D printing and why they can cause trouble to printed parts. Then, we moved on to the most common approach for handling undercut-related issues, using support structures.

Lastly, we discovered how layer orientation and part placement can change the mechanical behavior of our components, and we gave a few practical suggestions on how to optimize orientation choice.

Getting an overall idea of the covered subjects is super-important for the proper use of additive manufacturing (and, in particular, FDM printers).

Now that we have a better understanding of FDM printing's potential and its limitations, we can move on to the next chapter, where we are about to start our part setup and slicing journey!

16

Printing Our First Part

Now that we've covered the general theory behind FDM additive manufacturing, we can move on to Fusion 360 and its CAM module.

In this chapter, we will approach an example part, starting from the setup and then reviewing the whole process. The goal of this chapter is to provide you with all the tools needed to create a G-code file to export to a 3D printer.

In this chapter, we will cover the following topics:

- Presenting the model
- Creating a new printing setup
- Orienting the model onto the build platform
- Generating the support structures
- Simulating the toolpath
- Using the post-processor

Technical requirements

To understand this chapter, make sure that you have read and understood *Chapter 14* and *Chapter 15*, or have basic knowledge of 3D printing in general.

Presenting the model

Since we are about to start creating a printing setup using Fusion 360, we should begin with a 3D model to use as an example part. The part we are about to print is a little bit more advanced than the examples covered in *Chapter 15*. Here it is:

Figure 16.1: Example model

It's quite a complex bracket with several mounting holes and multiple section changes. This type of geometry may be a bit challenging for milling, but it is feasible for additive manufacturing.

First, we must check if it can fit inside the print volume of our printer. The overall dimensions of the part are 45x107x45 mm, and I plan to print the part using a printer with a build volume of 230x150x140 mm, so we shouldn't find any issues with the dimensions.

Now that we have discovered that we can print the part, we also need to understand the forces it will have to sustain. I can tell you that this component is not loaded with any force, so the layer orientation won't play an important role when choosing the printing orientation. Therefore, we can focus on choosing the part placement that minimizes the amount of support material needed.

The best placement is probably the one shown in the preceding figure since it will only require a tiny bit of support material to support the upper layers above the big hole.

Don't be too scared about part placement – if you don't immediately find the best solution to orient a part on the build platform, you can always rely on automatic part placement algorithms, which will do this for you!

We will cover these types of commands in the *Orienting the model onto the build platform* section.

Now that we have introduced the part we are about to print, we can launch Fusion 360 and dive into the **MANUFACTURE** workspace.

Creating a new printing setup

The first thing we shall do is move to the **ADDITIVE** panel inside the **MANUFACTURE** workspace, where we will find every 3D printing command to play with. Then, from the **SETUP** drop-down menu, select the **New Setup** command:

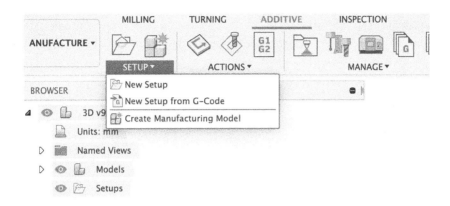

Figure 16.2: New Setup

The following panel is quite familiar and will be the same for any setup operation we face in terms of turning, milling, or cutting:

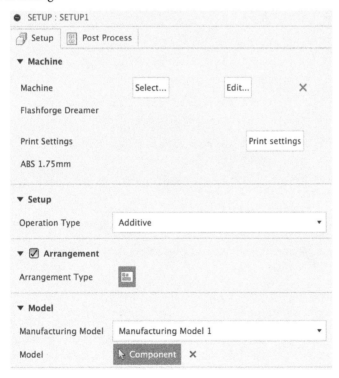

Figure 16.3: Setup's Setup tab

The first thing we must check is that **Operation Type** is set to **Additive**.

Next, we can pick the 3D printer we want to use. This is not a mandatory step, but since 3D printers usually have quite a limited printing area, we must ensure that the object we want to print doesn't exceed the maximum available printing dimensions. Choosing a machine will let us check its boundaries much easier.

To select a 3D printer, go to the **Machine** panel and click on **Select…**. The following window will appear:

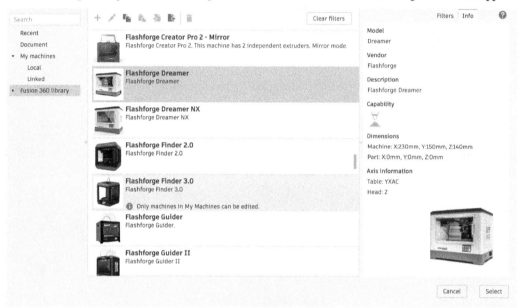

Figure 16.4: 3D printer library

Selecting a printer from the list lets us check its details on the **Info** panel on the right.

There is quite a large machine library for desktop FDM printers – I'm going to choose my trusty Flashforge Dreamer, which features a double extruder and a build volume of 230x150x140 mm.

Once we've picked our 3D printer, we must click on **Select** to get back to the **Setup** window from *Figure 16.3*.

Next, using the **Model** selection tool, we can pick the 3D models we want to print; in our example, we picked the bracket shown in *Figure 16.1*.

After that, there's an important group of settings we must check, which can be reached using the **Print Settings** button. Here, we can specify the material we want to use for printing:

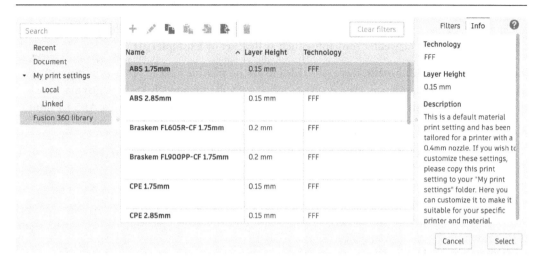

Figure 16.5: Print Settings

There are already many printing settings to choose from. For this example, we are going to use the default setting, which is named **ABS 1.75mm**. As you may have guessed, ABS is the material we are about to print, while 1.75 mm is the diameter of the filament printable by our 3D printer. If your 3D printer features a bigger extruder, you can go for **ABS 2.85mm**, which is the other most common filament diameter.

> **Note**
>
> In *Chapter 17*, we will dive much deeper into the printing settings, so we can leave the default settings as-is for now.

Now that we've covered the most important options for the **Setup** tab, we can go to the **Post Process** tab. Here, we can create a custom name for the generated G-code file or add a comment:

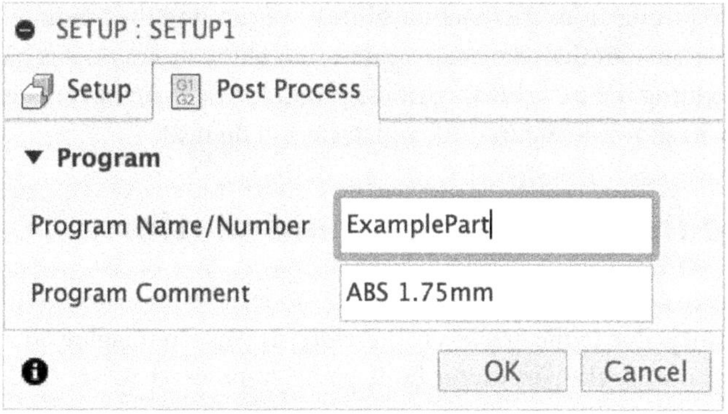

Figure 16.6: Setting's Post Process tab

I would suggest calling the G-code file something that helps us remember the part's shape. For example, for **Program Name**, we may enter something such as **Bracket** or **ExamplePart**.

When using **Program Comment**, it is a good idea to always provide a recap of the printing settings or the 3D printer model; this way, even in a year or two, we will immediately know which machine is intended to be printed onto. Since I only have one FDM printer, I'll go for a comment like **ABS 1.75mm**.

Now that the setup is complete, we can click on **OK** and look at the 3D environment:

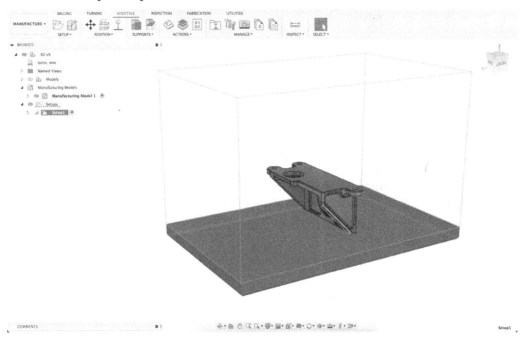

Figure 16.7: Build volume

Since we selected a 3D printer from the machine library, we can find the boundaries of the printing volume. The part is way smaller than the build volume, but at the moment, it is badly oriented onto the build platform; printing the part this way would require tons of support material. We must flip the part upside down, as shown in *Figure 16.1*, which we will do next.

Orienting the model onto the build platform

The contact area between the build platform and the 3D model is so important. We can safely say that 90% of a successful print is due to the first layer. Therefore, as you can imagine, there is a whole set of commands to orient and place the part on the build platform however we see fit.

Alongside the standard **Move Components** command we should already be familiar with, other specific tools in the **POSITION** menu are worth discussing:

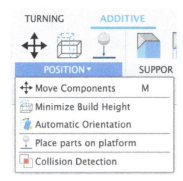

Figure 16.8: The POSITION menu's commands

Let's review them one by one.

Place parts on platform

The **Place parts on platform** command allows us to put every component we plan to print onto the build surface. For the example we are studying, this is not needed (since there is only one part and it is automatically oriented onto the build plate), but when printing multiple parts, we may have to use it.

Forget our example bracket for a moment – let's focus on another example:

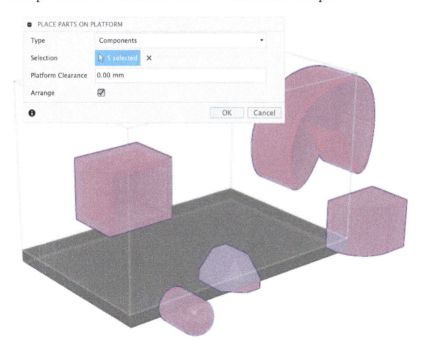

Figure 16.9: Place parts on platform (before)

As we can see, when there are multiple parts to print, they may not fit inside the printing volume as they are positioned in the 3D world. But by using the **Selection** tool and flagging the **Arrange** option (and then clicking **OK**), we can change their position and put them onto the build platform (this is something like a 3D nesting operation!).

After running the command, all the parts will be inside the build volume and touching the build surface:

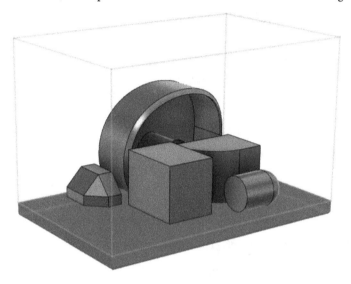

Figure 16.10: Place parts on platform (after)

Please note that this command doesn't rotate the parts – it simply moves them. Therefore, we shall always set their printing orientation first. So, let's find a couple of ways of changing the part orientation.

Minimize Build Height

The first and most simple command to orientate a part onto the build platform is called **Minimize Build Height**. It is a positioning command whose aim is to reduce the overall Z dimension of the printed part.

Let's see its behavior with the example part:

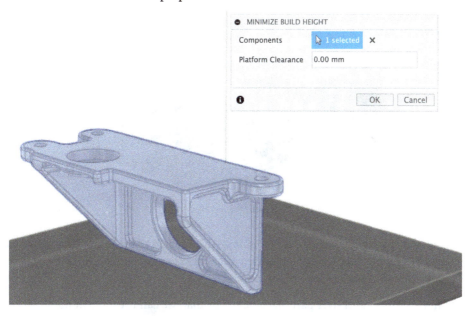

Figure 16.11: Minimize Build Height (before)

This is a rather simple command with just two options:

- **Components**: This lets us choose the part we want to rotate to minimize the build height.
- **Platform clearance**: This is a height value offset that raises the bottom face of the part above the build platform. Since the first layer is always touching the build platform 99.9% of the time, we can leave this field set to **0.00mm**.

Let's see the resulting orientation after using this command:

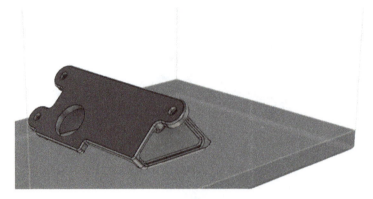

Figure 16.12: Minimize Build Height (after)

Fusion 360 calculated the best orientation to minimize the Z height of the printed part and rotated the component by 45°. Still, this is not an optimized orientation to print our part since this placement requires a lot of support material.

We should have more control over part placement. For that, we can use a much more advanced command called **Automatic Orientation**.

Automatic Orientation

Automatic Orientation is an operation similar to sheet metal nesting. It is a recursive operation that aims to optimize part placement in the build volume:

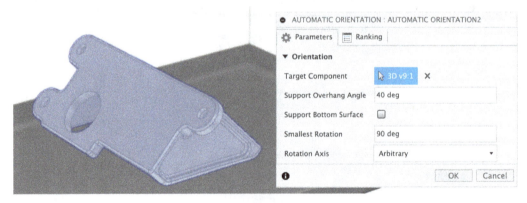

Figure 16.13: Automatic Orientation's Parameters tab

Automatic Orientation is quite a simple yet very powerful command, consisting of two tabs: **Parameters** and **Ranking**.

Parameters

Inside the **Parameters** tab, we can find a list of basic requirements for the part placement:

- **Target Component**: Using this selection tool, we can pick the 3D model we want to orient onto the build platform.

- **Support Overhang Angle**: This is the maximum undercut angle at which our 3D printer can print a sloped wall. This value is influenced by layer height, printed material, and nozzle diameter. However, most of the time, we can consider a maximum value of 40-45°. Therefore, we can leave the default setting at 40°.

> **Note**
>
> In 3D printing, sometimes, undercuts are referred to as **overhangs**. The meaning is the same: it's an area that is not supported by lower layers.

- **Support Bottom Surface**: This lets us specify that we want to support the first layer of material. This is quite odd since the build platform is always the best support surface available: we don't need support material if it is possible to let the part directly touch the build surface. Most of the time, we shall leave this box unticked.

- **Smallest Rotation**: This is the rotation angle step that will be used to find the optimal placement on the build platform. Much like when nesting a sheet metal part, smaller values will lead to a better solution but also a longer processing time. Most of the time, we can use a step angle of 90°. However, for complex parts, we may have to reduce this value by a few degrees. Since the example part is very simple, we can go for 90° or for 45°.

- **Rotation Axis**: This is the axis that's used to calculate the best part orientation. We shall always leave this value set to **Arbitrary**.

Now that have looked at the first tab, we can move on to the next one.

Ranking

Now that we've specified how the part can be rotated onto the build platform, Fusion 360 can start rotating the part until it finds the best orientation. However, we must still tell Fusion 360 how to recognize the best part placement. This can be achieved in the **Ranking** tab:

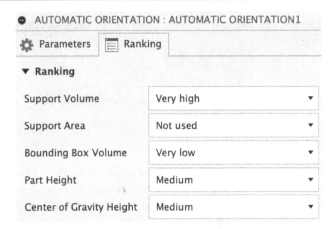

Figure 16.14: Automatic Orientation's Ranking tab

There are five different optimization targets here. Most of the time, we are going to use the default settings, but it is worth mentioning their meaning:

- **Support Volume**: This evaluates the amount of support material needed to print the part. A placement with no support material will reach a higher score, while a placement that requires tons of support material will get a lower rank. This is the most important target, so by default, it has a **Very high** value.

- **Support Area**: This measures the build area used by support structures. A larger area will lead to a lower score, while a smaller area will lead to a higher one. Most of the time, this is not an interesting parameter since we mostly care about support volume. Therefore, by default, this ranking is **Not used**.

- **Bounding Box Volume**: This evaluates the overall dimension of the bounding box that contains the part along the x, y, and z axes. Once again, this is not a useful ranking value, so by default, it is set to **Very low**.

- **Part Height**: This target gives a higher score to placements that orientate the part in a manner that reduces the print height. For example, a tall box will have a higher score when laid on its sides rather than standing up. This is quite an important ranking since a taller part may be influenced more by vibrations and thermal distortions (as we will discover in *Chapter 17*). By default, it has a value of **Medium**.

- **Center of Gravity Height**: This gives a higher score to part placements with a center of mass closer to the build platform and a lower score to placements with a higher center of mass. This setting has a default value of **Medium**.

I know that the **Automatic Orientation** command may sound a bit complex, but let's recap its intended usage. First, we specified how the part can be rotated onto the build platform, the maximum overhang angle that doesn't need support structures, and the angle step to find the best orientation.

Then, we specified a set of rankings that will let Fusion 360 decide which is the best part placement. In our case, part placements will be ranked mostly by support material usage.

Fusion 360 will then judge the part placement that uses the least amount of support material and will propose that placement as the best solution.

Let's check the results:

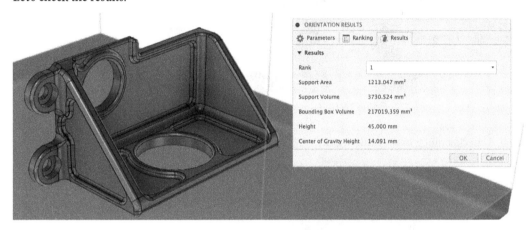

Figure 16.15: Automatic Orientation – Rank 1

After running the command, you may have noticed that a new tab called **Results** appeared. Here, we can find all the rotation steps that Fusion 360 performed to find the best part placement.

From the **Rank** drop-down menu, we can select every resulting orientation sorted by the rank result.

For now, we can look at **Rank 1**, meaning that the placement displayed in the preceding figure is the best one according to Fusion 360, with a **Support Volume** of **3730 mm³**.

Let's look at some other ranks. **Rank 2** is the part orientation I plan to use for the printing setup; it requires a slightly higher amount of support material (**3854 mm³**) than the previous case, but I think it's much easier to print (both are good options anyway):

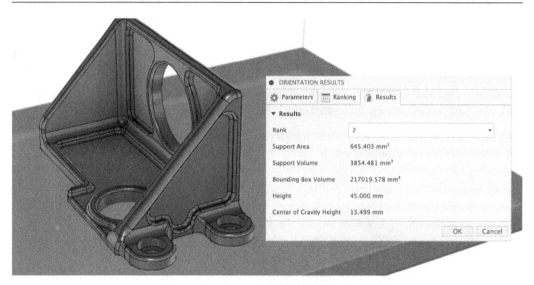

Figure 16.16: Automatic Orientation – Rank 2

Let's also check a much lower rank, **Rank 7**:

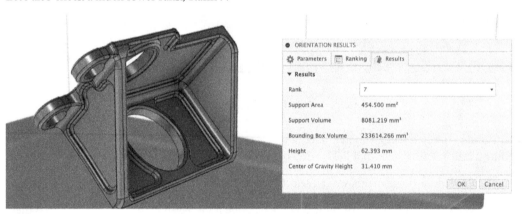

Figure 16.17: Automatic Orientation – Rank 7

This is one of the multiple rotation steps that Fusion 360 evaluated as a possible part placement. It's easy to see why it has such a low ranking:

- It has a very small contact area with the build surface
- It uses more than double the amount of support material used by Rank 1 and Rank 2
- It has a significantly higher center of mass
- It requires a much taller print

Please note that automatic part orientation doesn't consider the ease of printability or the mechanical properties of the resulting object. That's why we sometimes have to go for a lower-ranking placement.

Now that we've studied part placement, we can move on with the printing setup and dive into support structures.

Generating the support structures

Two commands generate support structures in Fusion 360, and they can both be found in the **SUPPORTS** dropdown menu under the **ADDITIVE** tab. The first and most common support command is called **Solid Volume Support**, while the second is **Solid Bar Support**:

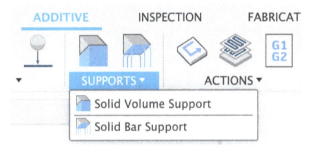

Figure 16.18: SUPPORTS commands

Straight off the bat, I can tell you that **Solid Bar Support** has very poor performance on most FDM printers (it is widely used for SLA printers instead). Since SLA printers have limited support in Fusion 360 and since we are printing an FDM part, we will focus on **Solid Volume Support** only. Once launched, you'll see this panel:

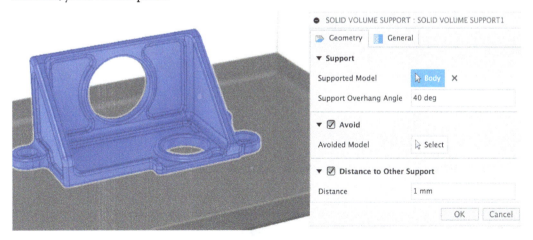

Figure 16.19: Solid Volume's Geometry tab

This command is not complex to use, with just two tabs. Let's look at the first one – the **Geometry** tab:

- **Supported Model**: Using this selection tool, we can specify the 3D model we want to create support material for.

- **Support Overhang Angle**: This is the maximum overhang angle that doesn't need support structures. This value is influenced by layer height and nozzle diameter. However, most of the time, a value of 40° will work just fine.

- **Avoid**: This option lets us specify geometries (faces or parts) that we don't want to be supported by any material. For our example, we can untick this option as there aren't any structures to avoid.

- **Distance to Other Support**: This is the minimum distance that new support structures can be from other existing support structures. Once again, we can safely untick this option for the example model.

> **Note**
>
> Support structures created with FDM printers may be difficult to remove once the print is complete. This is especially true if they are very dense and near fragile geometries. That's why we sometimes have to limit their generation. We will learn how to manage support generation in more detail in *Chapter 17*.

Now, let's move on to the **General** tab:

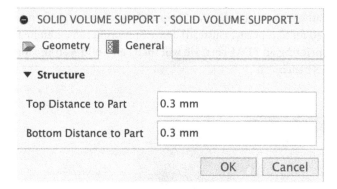

Figure 16.20: Solid Volume's General tab

Top Distance to Part and **Bottom Distance to Part** are two similar parameters and are used to reduce the bond between the support structures and the printed part.

They reduce the bond strength by adding a small gap between the support material and the printed geometry. This way, once the print is complete, it is easier to remove the support material:

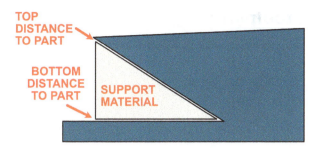

Figure 16.21: Support distance to part

These gaps depend on the layer height and the printing material. We may have to fine-tune them on our printer via some trial and error. The default value of **0.3 mm** will likely work just fine in most scenarios.

Now that we've reviewed how to create support structures, we can check the resulting model:

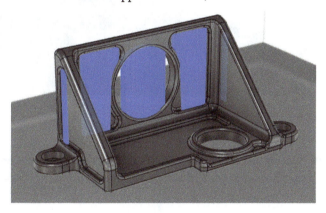

Figure 16.22: Generated support structures

The blue geometries are the generated support structures. Through a preliminary check, we can see that we don't have any unsupported overhang geometries anymore!

And that's all we need to know about support material generation. We can now move forward and check the generated toolpath!

Simulating the toolpath

At the moment, our screen should look a bit like this:

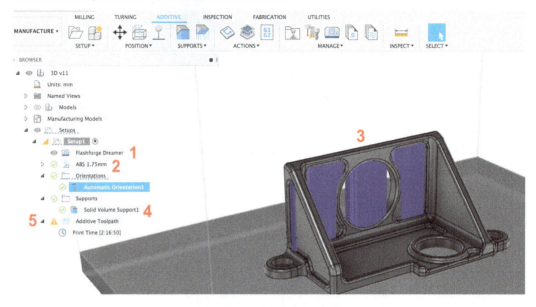

Figure 16.23: Printing environment

The 3D printer (and its printing volume) have been set (*1*), the printing settings have been selected (*2*), the part has been properly oriented onto the build platform (*3*), and the support structures are up to date (*4*).

However, there is still a warning for the **Additive Toolpath** area (*5*) – the yellow triangle on the left means that the toolpath data is outdated.

Luckily, updating the toolpath is very simple. To do this, we just need to click the **Generate** command, which we can find in the **ADDITIVE** tab under the **ACTIONS** drop-down menu:

Figure 16.24: The Generate command

After updating the toolpath, we can check the simulation results using the **Simulate Additive Toolpath** command:

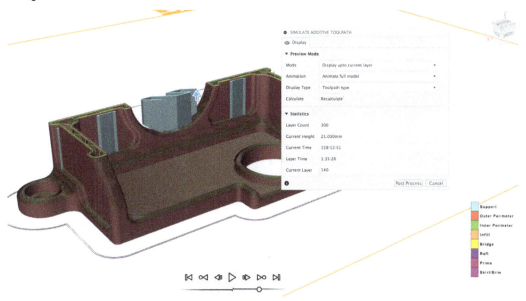

Figure 16.25: Simulated toolpath

The simulation environment is quite similar to those found for milling and turning – there is a timeline at the bottom and a set of play, stop, and rewind buttons to control the simulation's progress.

You may have noticed a vertical arrow. This allows you to analyze the printed part layer by layer. This way, it is possible to check the infill layers or the support material section to find any potential issues for surface finish or part strength. Let's look at a typical example where checking the simulated toolpath can prevent us from printing a weak geometry:

Figure 16.26: Simulated toolpath – layer 23

Here, we can find **Layer 23** and the infill pattern generated. I think that the resulting strength of the part may be a bit weak, so we may need to increase the infill percentage by tweaking the printing settings. This is exactly what we are going to do in the next chapter.

For now, let's suppose we are fine with the current results, and we want to export the file to our 3D printer. As we may remember from what we learned for milling and turning, to convert CAM operations into G-code, we need to use a post-processor.

Using the post-processor

Now that we have completed the setup process and simulated the printing process, we can safely generate the G-code for our 3D printer via a post-processor.

As you may recall, a post-processor is a program whose scope is translating CAM operations into G-code files. Fusion 360 has quite a large set of post-processors for most of the desktop 3D printers on the market.

Let's start the command. You can find the **Post Process** command in the **ADDITIVE** tab under the **ACTIONS** drop-down menu:

Figure 16.27: The Post Process command

This will launch the following **Post Process** window:

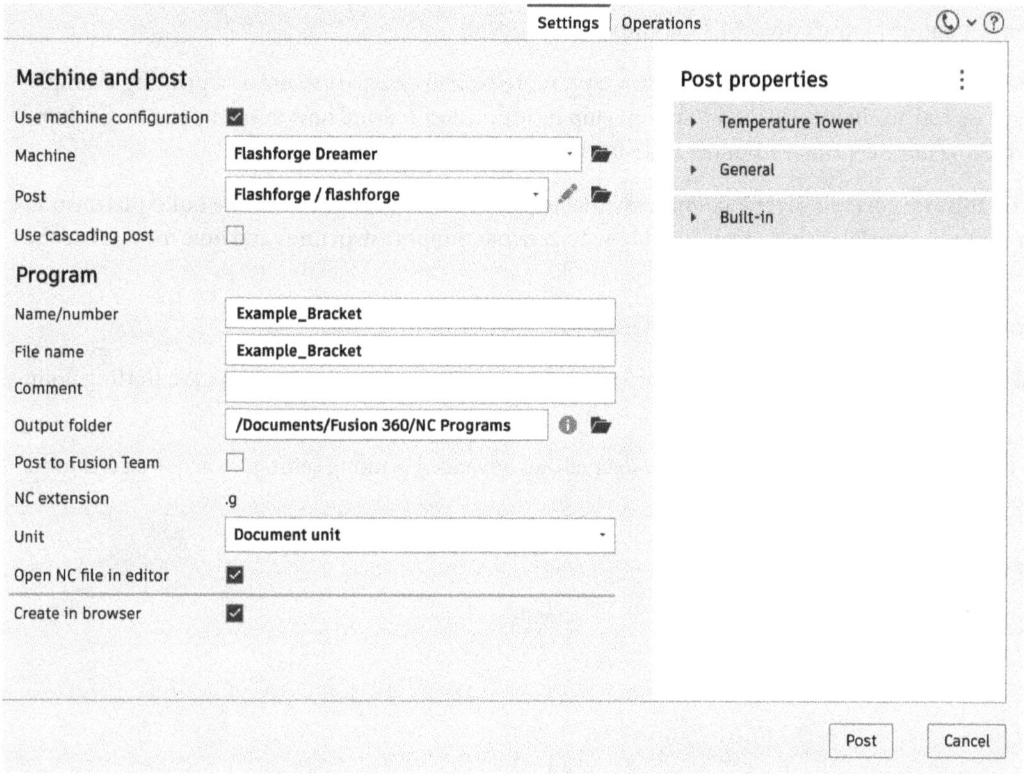

Figure 16.28: The Post Process window

The only option we will care about here is the machine model. We can browse all supported 3D printers from the **Machine** drop-down menu (I picked **Flashforge Dreamer**).

If our printer is supported by Fusion 360, there is not much else we have to do here as the default settings will work just fine; we just have to click on **Post** and export the generated G-code file to the 3D printer.

> **Note**
>
> If your 3D printer is not supported by Fusion 360, there is no easy workaround. My only suggestion is to find a post-processor (for another printer) that is fairly compatible with your printer as well. Then, you can start exporting the same toolpath with several post-processors and try to print the generated G-code files. Chances are high that, between all those files, one of them will work just fine.

That's all we need to know to start using Fusion 360 for basic FDM printing processes.

Summary

Congratulations – you've reached the end of this chapter!

Let's recap what we went through. First, we presented a real case part to use as a printing example. Then, we learned how to create a printing setup inside Fusion 360 and how to set the printing volume by picking our 3D printer from the machine library.

After this, we covered several commands that let us position the parts onto the build platform in the best way possible. Then, we learned how to generate support structures and how to simulate the generated toolpath.

Lastly, we learned how to generate the G-code to send to the 3D printer.

All these steps represent the typical approach to any FDM printing process and are the starting point for more advanced operations.

In the next chapter, we are going to dive deeper into advanced printing settings to acquire even more tools for complex parts.

17
Understanding Advanced Printing Settings

Most of the time, the default print settings are more than enough to get good results for our objects. However, sometimes, for example, when printing exotic materials or when experimenting with complex parts, we may have to tune them a little.

Unfortunately, additive manufacturing is a rather complex process, and FDM printers have a lot of settings we must at least understand in order to master the technology.

In this chapter, we will dive straight into printing settings, reviewing the most important ones that every beginner should learn. The goal is to show, as much as possible, how different settings can influence the printing results.

So, in this chapter, we will cover the following topics:

- Creating a new printing preset
- Understanding general parameters
- Understanding extruder parameters
- Understanding shell parameters
- Understanding infill parameters
- Understanding print bed adhesion
- Understanding support material parameters
- Understanding speed parameters
- Understanding tessellation parameters

Technical requirements

We are about to cover many settings related to the entire FDM printing process. Therefore, as you can guess, this is quite an advanced chapter; having read *Chapters 14, 15*, and *16* is a mandatory starting point.

Creating a new printing preset

All the options we are about to cover can be accessed when editing the printing profile assigned to an additive manufacturing setup. To reach the options, we just have to expand **Setup1**, under the design tree, and right-click on the printing profile, which in the example is called **ABS 1.75mm**.

Figure 17.1: Printing profile

In the menu, click on **Edit** to access all the printing settings contained inside **Print Setting Editor**:

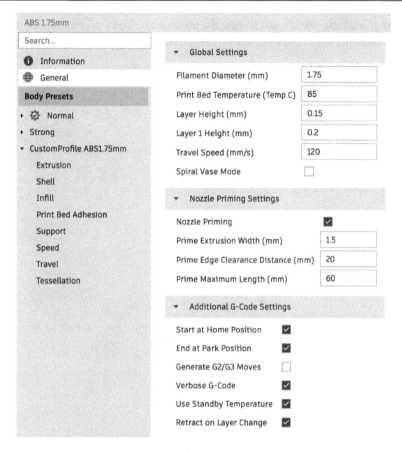

Figure 17.2: Print Setting Editor

On the left, we can find all the printing presets for the selected material (in our case, **ABS 1.75mm**). At the moment, there are two built-in printing presets – **Normal** and **Strong**. The main difference between the two is the density of the material interior and other minor details, such as the number of perimeters and the number of top and bottom layers.

We can duplicate a default preset to create a custom printing profile. To do this, we just have to right-click on an existing profile and select **Duplicate**:

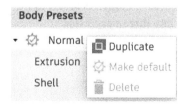

Figure 17.3: Duplicating profiles

After duplicating an existing profile, we can give it a preset name and a preset description. Here, we shall type something that may help us in the future to remember what changes were made to the default set of parameters.

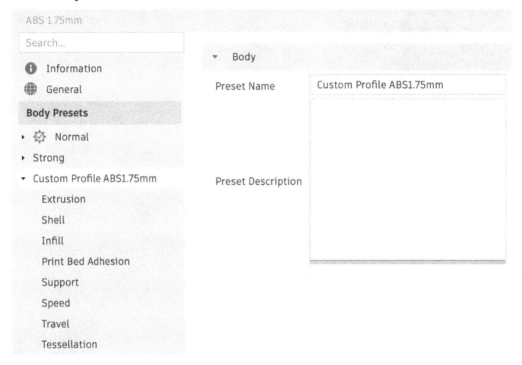

Figure 17.4: Custom profile

Now that we have created a custom printing preset, we can focus on its parameters and their influence on the printing process.

Please note that every section in this chapter will refer to the panels found in *Figure 17.2* on the left, just below **Body Presets**.

Understanding general parameters

In this section, we will review the general parameters found in the **General** tab that you can see in *Figure 17.2*. The options are divided into three categories:

- **Global Settings**
- **Nozzle Priming Settings**
- **Additional G-Code Settings**

The most important settings are contained inside the **Global Settings** section, while the **Nozzle Priming Settings** and **Additional G-Code Settings** options are way too advanced for beginners.

Let's look at the most important **Global Settings** options we have to learn (as a rule of thumb, we shouldn't mess with parameters unless we are absolutely confident of their usage, as they can ruin our prints):

- **Filament Diameter (mm)**: This is the diameter of our 3D printing filament. Most 3D printers on the market today extrude filaments with a diameter of 1.75 mm or 3 mm. If we want to be sure about the diameter, we can measure the filament using a caliper on multiple points. After measuring the filament, we may discover that it is a bit bigger or smaller than the nominal value, so we can now insert the proper size; for example, the size may be something such as 1.73 mm or 1.78 mm.

- **Print Bed Temperature (Temp C)**: This value is the temperature of the 3D printer's heated bed, which has to be set following the specifications provided by the filament's manufacturer. Depending on the type of material, we may need to disable the heated bed or increase its temperature up to 120°C. For example, most ABS filaments require a heated bed at around 105-110°C. We should set the temperature accordingly.

- **Layer Height (mm)**: This parameter is the height of every printed layer; smaller layers ensure a higher resolution and a smoother surface quality, at the cost of higher print times. The typical layer height largely depends on the nozzle diameter; however, since a nozzle of 0.4 mm is almost a standard for every printer on the market, we can set **Layer Height** of 0.1 mm for a nice smooth surface finish, up to 0.3 mm for a coarse surface finish. In the following figure, we can see the finish difference between a very thin layer and a thicker one.

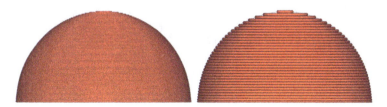

Figure 17.5: Layer Height

Here, the sphere on the left has a much thinner layer height, while the sphere on the right is much coarser. Please note that on most printers, decreasing **Layer Height** to below 0.1 mm will not give better results; that's why 0.1 mm is somewhat of a lower limit to the layer thickness.

- **Layer 1 Height (mm)**: This is the height of the first layer printed on the build platform. Most of the time, it is suggested to increase this thickness to improve layer adhesion to the platform and to prevent the nozzle from rubbing against the heated bed.

- **Travel Speed (mm/s)**: This is the travel speed of the extruder when not printing. It largely depends on your machine, including how fast it can move and how much it vibrates.

- **Spiral Vase Mode**: Long story short, enabling this option will force the extruder to print only the outer contour of our part in a long spiral. To get a better idea of the effect, let's check the following example:

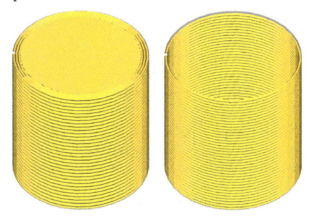

Figure 17.6: Spiral Vase Mode

On the left, there is a conventional printing toolpath, while on the right, **Spiral Vase Mode** was turned on; as we can see in the figure, the cylinder is now hollow and only the outer contour has been printed.

That was all we had to know about the general printing settings. Let's focus on the next set of parameters related to the extruder.

Understanding extruder parameters

In this section, we will review all the parameters related to the extruder, its temperature, and its behavior when printing:

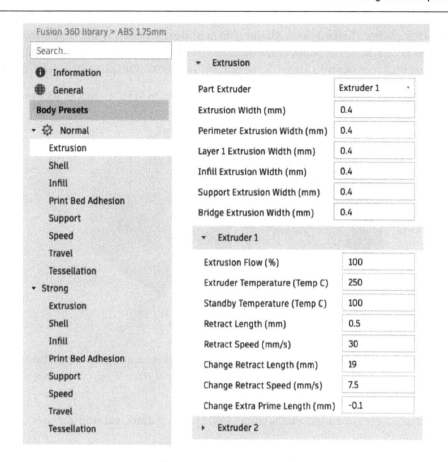

Figure 17.7: Extrusion panel

The options are divided into two main categories:

- **Extrusion**: A list of settings that control the extrusion width of every extruder
- **Extruder #**: A list of options to control each extruder, independently controlling parameters such as temperature and extrusion flow

Let's take a look at these in more detail.

Extrusion options

The following parameters are mostly related to the dimension of the nozzle and are rarely changed:

- **Part Extruder**: When using a multi-extruder printer, this selection panel can specify which extruder shall be used to print the part geometries.

- All the options from **Extrusion Width (mm)** up to **Bridge Extrusion Width (mm)**: These values are used to determine the width of the toolpath. As a best practice for most applications, the extrusion width can be set to a range between 100% and 200% of the size of the nozzle diameter.

  So, for example, if our printer has a nozzle with a diameter of 0.4 mm, we can set an extrusion width from 0.4 mm up to 0.8 mm.

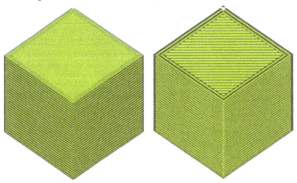

Figure 17.8: Extrusion Width

Please note that **Layer Height** is equal in both examples of the preceding figure, but on the left, since the line width is smaller, lower printing widths result in a better surface finish (this is especially true for complex geometries) and better overall precision.

On the other hand, higher extrusion widths ensure lower printing times and stronger layer adhesion.

> **Note**
> Why does a higher extrusion width decrease printing times? The reason is simple: since the lines are wider, they can cover the same area with a smaller number of passes!

There is no magic rule to set the extrusion width – my suggestion is to start using a width equal to the nozzle diameter and change the values if needed.

Extruder 1 options

The following parameters are related to the printing material extruded by **Extruder 1**, but it is possible to adjust printing settings individually for each extruder if needed:

- **Extrusion Flow (%)**: This is a multiplier that overrides the amount of material extruded. We may have to adjust this parameter if the width of a printed line is different than expected.

  As you can see in the following figure, when increasing the flow, there is more material extruded through the nozzle than needed; therefore, the resulting printing width will be higher:

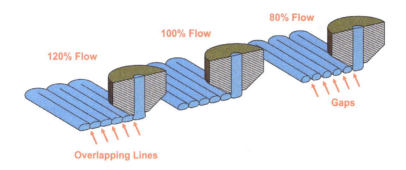

Figure 17.9: Extrusion Flow multiplier

Since this higher amount of material creates overlapping areas between the passes, increasing the material flow grants a higher layer strength. In contrast, decreasing the flow will create gaps between the extruded lines since there is not enough material to fill the space. Ideally, we should tweak this parameter only if the nominal extrusion width is different from reality.

- **Extruder Temperature (Temp C)**: This is one of the most important parameters to keep in mind – it is the softening point of our printed polymer. Every material has a different extruding temperature, so we shall always refer to the temperature suggested by the polymer supplier; for example, PLA may be extruded at around 180°C while ABS has to be extruded at around 220°C.

  Extruding a polymer at the wrong temperature will likely clog the extruder or result in a bad-quality print – we must always check the temperatures!

- **Standby Temperature (Temp C)**: This is a parameter mostly used for multi-extruder printers. It is the temperature of the extruders that are not printing; once these extruders have to start printing, they will be heated up to the extruding temperature and then start printing.

- **Retract Length (mm)**: This value measures how much the filament shall be retracted when the extruder is not printing. Higher values are needed for very soft materials, while smaller values can be used for denser materials.

- **Retract Speed (mm/s)**: This parameter measures how fast the filament is retracted when the extruder is not printing. Smaller values should be used for stiffer materials, while higher ones should be used for softer materials.

Those were the most important options to master the extrusion parameters; we can now move on to the next section, where we will learn about printed geometries.

Understanding shell parameters

In this section, we will review all the options that let us control the outer geometries of our part. This is very important since they affect what you can actually see once the part is printed.

As we can see in the following figure, there are three different types of outer geometries to consider:

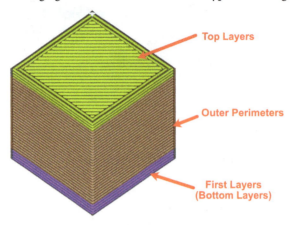

Figure 17.10: Shell surfaces

The options we are about to cover handle all these different surfaces:

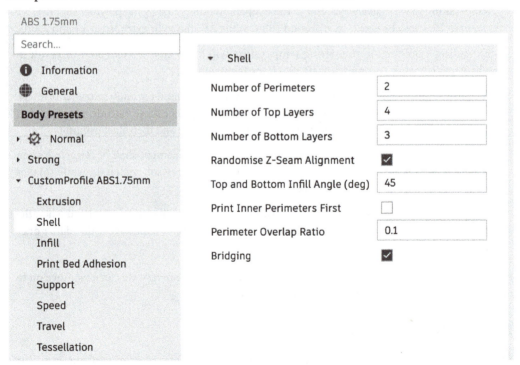

Figure 17.11: Shell panel

Let's break down the options:

- **Number of Perimeters**: This is the number of perimeters to be printed in the outer contour. As you can imagine, more perimeters take longer to be printed but they also help create a stronger part (this is especially true for complex shapes).

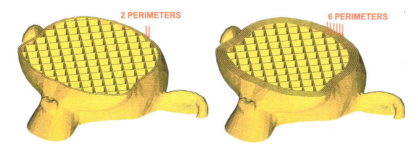

Figure 17.12: Outer perimeters

On the left of the preceding figure, we can see a rather typical print with two perimeters, while on the right, we have a pretty sturdy print with six outer perimeters for a very tough part. Most of the time, we can go for two to three perimeters only. A much more effective way of increasing the part strength is increasing the density of the infill geometry. We will discover how to do just that in the next section.

- **Number of Top Layers**: This is the number of layers to be printed on top of the part. We will increase the number of top layers especially if the infill density is low. This can be a bit tricky to understand, so let's take a look at the following example:

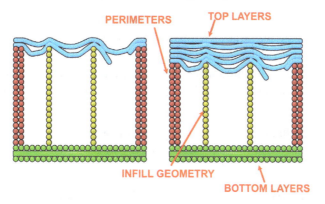

Figure 17.13: Top layer infill

The figure displays two parts – both feature a very low infill density; however, the part on the left features only two top layers, while the part on the right has five of them. As we can see, the extruder had problems when printing the top layers of both parts: the material is not supported enough. But the part on the right still has a good top surface finish since the last layer could be printed on the adjacent top layers previously printed.

- **Number of Bottom Layers**: This is the number of bottom layers printed. We should always print at least two or three bottom layers; otherwise, the part will likely get damaged while detaching it from the build platform.

- **Randomize Z- Seam Alignment**: This option controls the location around the perimeter of the part where the extruder will start and end. This is important because typically, the starting point of each layer has tiny defects:

Figure 17.14: Randomize Z-Seam Alignment

The light dots on the parts display the starting points for every layer; in those locations, it is likely to have a tiny bump if the printing settings are not properly set. On the left of the preceding figure, we can see the default behavior of Fusion 360, where every entry point is aligned, and on the right, we have enabled **Randomize Z-Seam Alignment**, which evenly distributes the dots around the part. Please note that random entry and exit points can also help with increasing the part's strength.

- **Perimeter Overlap Ratio**: This option lets us specify how much each perimeter should overlap with another. High overlap values ensure a stronger part at the expense of worse precision and surface quality.

Now that we have a clear idea of what is going on with the outer surfaces of the print, we can focus on the inner structures.

Understanding infill parameters

In this section, we will cover all the options to manage the inner areas of our part – in particular, the infill geometries and their specifications:

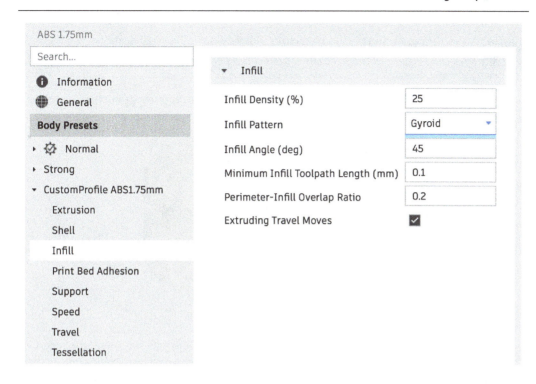

Figure 17.15: Infill panel

Let's look at the options:

- **Infill Density** (%): This is the amount of infill material printed. A value of 100% will result in a solid part, while a value of 0% is an empty shell. Usually, this value is set around 30%, depending on the loads applied on the part.

   As we can see in the following figure, changing the infill density can drastically change the part weight and strength as well as the printing time:

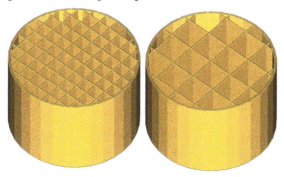

Figure 17.16: Infill percentage

The part on the right features infill geometries with a density of 10%, while the part on the left has an infill density of just 20%. This means that the inner structures of the part on the left are twice as strong, but also twice as heavy and take twice as long to print.

- **Infill Pattern**: This is the shape of the infill material. There are multiple geometries to choose from; this includes **Gyroid**, **Honeycomb**, and **Squares**, which are very popular with users:

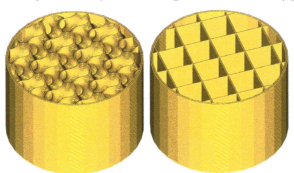

Figure 17.17: Gyroid versus Squares infill

On the left, we can see an example of the **Gyroid** infill, while on the right, we can see the **Squares** infill. My suggestion is to choose an infill pattern that avoids sharp turns when printing the structures, as this will greatly reduce vibration due to sudden speed changes… that's why I love the **Gyroid** infill pattern so much!

- **Perimeter-Infill Overlap Ratio**: This value specifies how much the infill geometries should overlap the layer perimeter. Higher overlapping values will result in a stronger part, but also in a worse surface finish with less precision.

That's all we need to know about infill patterns – there are some minor options that we left out, but to tell you the truth, it is very rare to deal with them; therefore, we can forget about them.

Now that we have looked at all the most important options to master infill geometries, we can move on to another hot topic: bed adhesion.

Understanding print bed adhesion

I was unsure whether to include this section or not, as it is a bit advanced for a beginner who wants to jump straight into printing, but on the other hand, it can be a lifesaver for many real-world scenarios.

The core of this section is how to improve the bond between the print bed and our part. As you may expect, if the printed part partially or fully detaches from the build platform while being printed, it becomes garbage in no time.

You may be wondering what may cause part detachments. The reason is pretty simple: thermal expansion (or shrinkage)! That's why we have to understand the temperature distributions inside a printed part a bit better.

Thermal expansion and shrinkage

There is a key concept behind all the options we are about to discuss: every material on the planet shrinks and expands when cooling or heating occurs.

When printing a part, we never have a constant temperature on our part:

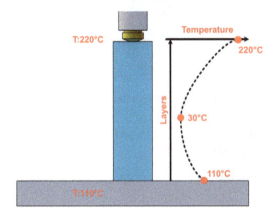

Figure 17.18: Temperature profile

As we can see here, there are two main sources of heat: the heated build plate and the extruding nozzle. The extruder has a temperature of 220°C, while the heated bed has a temperature of 110°C. This is a delta of 110°C.

To make things even worse, the layers in the middle of a tall part are quite far from both the extruder and the build plate; therefore, they will cool down almost to room temperature (if our printer doesn't feature an insulated build chamber).

A temperature differential of 190°C is prone to create temperature tensions between the layers. The different shrinking ratios of different layers can cause cracks and detachments from the build platform.

The typical issue found when printing materials such as ABS is corners detaching from the build plate:

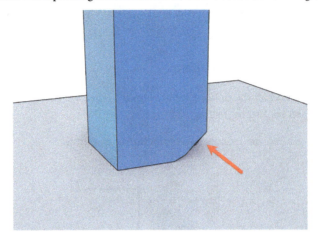

Figure 17.19: Corner detachment

If our prints feature an issue similar to the one shown in the preceding figure, we most likely have a problem with temperatures and cooling.

This was a long prelude to introduce a specific set of options that can handle this type of issue. Let's start dealing with them.

Print bed adhesion

There are three different techniques related to print adhesion that Fusion 360 offers:

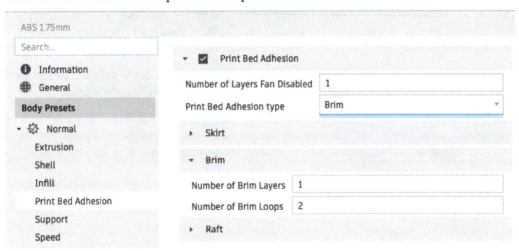

Figure 17.20: Print Bed Adhesion

These techniques are as follows:

- **Skirt**

- **Brim**

- **Raft**

Don't worry, we don't have to study all of them. First of all, I'd like to point out that **Skirt** shouldn't be included in this section since it is not a way to improve bed adhesion; it is rather a priming technique for the extruder. Therefore, we won't cover this option. **Raft**, on the other hand, is quite an advanced adhesion technique that, most of the time, is a bit overkill for common parts; therefore, we won't cover it either.

So, let's look at the remaining option, **Brim**, which is the most commonly used technique and easy to understand.

Brim is a way of printing a certain number of perimeters around the first layer; this additional material will increase the adhesion of the first layer to the platform.

Figure 17.21: Brim

In this example, the part on the left has **Print Bed Adhesion type** set to **Brim**; therefore, as we can immediately spot, it has a large number of perimeters printed around the bottom layer. This technique is particularly useful if the bottom layer has many sharp corners.

Let's now focus on the main options to control **Brim**:

- **Number of Brim Layers**: With this option, we can specify how thick the brim support should be. 99% of the time, we can leave this value at 1.

- **Number of Brim Loops**: This option lets us specify how many perimeters we want to print around the first layer. This number depends on part complexity. Most of the time, we can get successful prints with 3-5 contours.

As always, we shall take every option with a grain of salt and ask ourselves what the best printing strategy for the given geometry is.

For example, in the preceding figure, we can see two very simple geometries without sharp corners on the first layer. Therefore, unless we want to print with exotic materials with a crazy-high temperature expansion coefficient, we can most likely disable brim generation and go for the print shown on the right without any additional adhesion geometries.

That was all a beginner must know about adhesion techniques; it is now time to move on to looking at support material!

Understanding support material parameters

In this section, we are going to study the most important parameters related to support material and its shape:

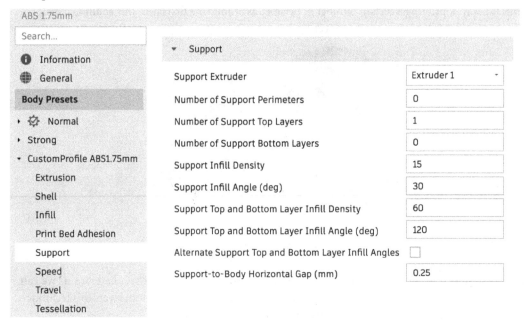

Figure 17.22: Support panel

Let's look at the options:

- **Support Extruder**: Using this option, we can specify which extruder shall print the support structures. This only applies if our printer has more than one extruder.

- **Support Infill Density**: The name is self-explanatory – this parameter is the density of the generated support structures. A higher density means a larger contact area with the part and stronger support.

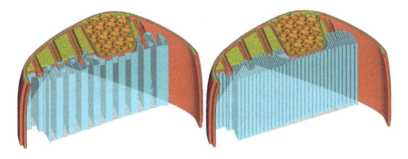

Figure 17.23: Support density

In the preceding figure, we can see a rather complex part being printed with support structures: on the left, we can see a support density of 15%, while on the right, the density is 50%. As we can guess, the part on the right will greatly benefit from the higher support density for all overhang geometries.

Now that we have reviewed the most important settings to generate support structures, we can move on to the next section, where we will discover speed and acceleration parameters.

Understanding speed parameters

In this section, we will cover the most important parameters that control printing movements. This set of parameters is really important to understand and manage since wrong values can lead to vibrations and motor overheating, to name just a couple of typical issues.

As we can see in the following figure, there are many values we can set inside the **Speed** panel:

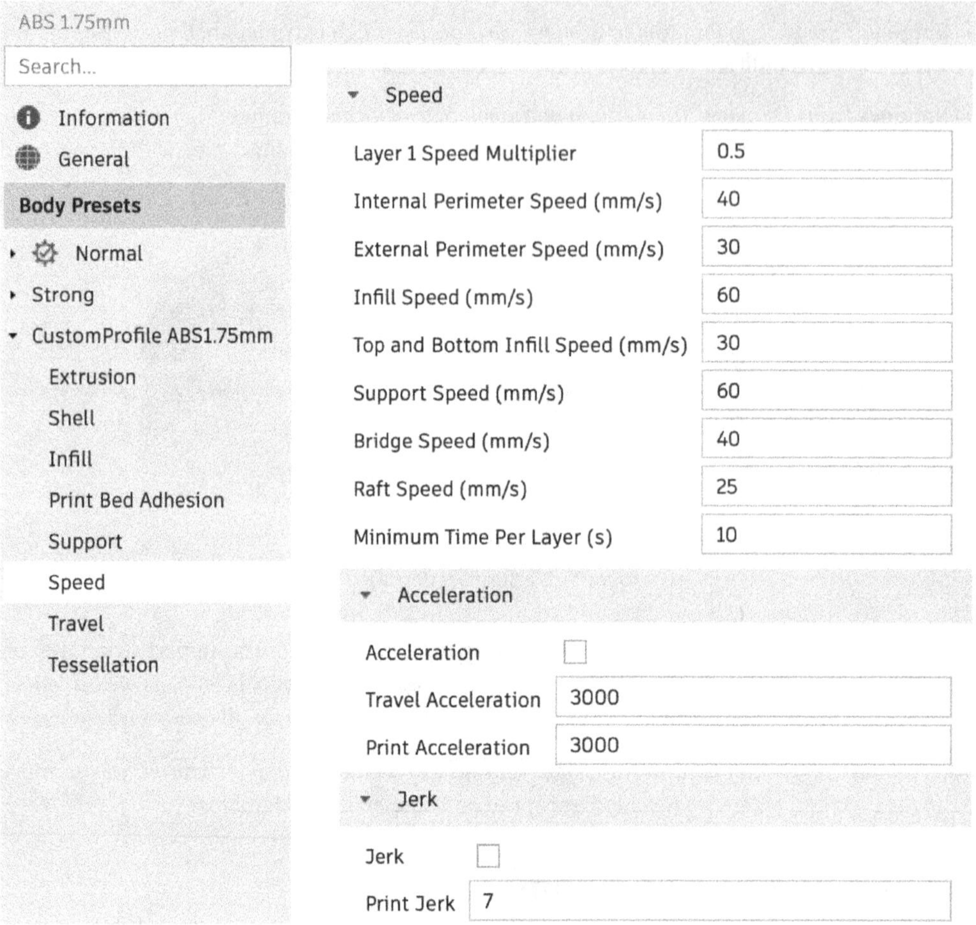

Figure 17.24: Speed panel

Please note that I cannot give you a startup value for all these parameters, since they are highly dependent on the machine's performance; sturdier and more powerful machines can print much faster than entry-level ones. My suggestion is to check the default speed values supplied by your machine brand and stick to them.

The key idea behind this set of parameters is to set movement limitations on our machine, a bit like the speed limits we find on the roads every day. Even if our machine can move much faster, these parameters are intended to preserve machine life and improve print quality.

As we are about to discover, not only can we control speed, but we also have much greater control over the printing movements. Let's review the options in detail.

Speed

Speed control is very intuitive. With the following parameters, we can control how fast the 3D printer should move while working:

- **External Perimeter Speed (mm/s)**: This is the printing speed used by the extruder when printing the outermost contour. Since this contour determines the surface finish of the part, it is always suggested to reduce the printing speed as much as possible.

- **Internal Perimeter Speed (mm/s)**: This is the printing speed for all the other perimeters. Since these geometries are inside the part (and cannot be seen), it is possible to print them a bit faster than the outermost perimeters.

- **Infill Speed (mm/s)**: This is the printing speed for the infill geometries. Usually, since the infill pattern is quite simple and it is hidden inside the part, we can use, as the infill speed, the highest printing speed achievable by our machine when extruding the chosen material. Increasing the infill speed as much as possible can drastically reduce the overall printing time.

- **Layer 1 Speed Multiplier**: As we found out when talking about bed adhesion, the first layer is always the most important: we must guarantee perfect adhesion to the build platform. That's why, in order to have better stickiness, it is always recommended to reduce the printing speed as much as possible. Modifying this parameter, we can set a percentage of reduction from the default printing speed. A common speed multiplier for the first layer is 0.3 to 0.5.

- **Top and Bottom Infill Speed (mm/s)**: This is the printing speed for the first and last layers. Printing at reduced speeds can improve bed adhesion and the overall surface quality – that's why, most of the time, we use the same speed as used for the outermost perimeters.

- **Support Speed (mm/s)**: This is the printing speed when printing the support structures. Since the support pattern is always very simple, and since it is also disposable, we shall always print these structures as fast as possible with our machine.

Acceleration

Up to this point, we've discussed the most important speed values to set; however, advanced printers can also control accelerations.

What is **acceleration**? If speed is how fast an object is moving, acceleration is how fast its speed is changing, and it is measured in mm/s^2.

Let's review the parameters related to acceleration:

- **Acceleration**: Flagging this option will set a command inside the generated G-code file to manage the acceleration of the 3D printer. However, please note that not every 3D printer can control accelerations via G-code.

- **Travel Acceleration**: This is the maximum acceleration for rapid movements. A higher acceleration ensures a more responsive machine and faster prints; however, we may want to reduce the maximum acceleration a little bit in order to reduce vibrations and resonances. Also, the value must be equal to or lower than the maximum acceleration reachable by our printer, to fit within its mechanical limits.

- **Print Acceleration**: This is the maximum acceleration for printing movements.

Jerk

What is **jerk**? Basically, it is how fast the printer's acceleration changes, measured in mm/s^3.

Since not every machine is capable of controlling jerk, we should definitely check the specifications of our printer if we plan to use it.

Let's check the options available to control jerk:

- **Jerk**: This option enables jerk control.

- **Print Jerk**: This is the maximum jerk reached by our machine while printing. Higher values will reduce printing time at the cost of more vibrations and resonances. If while printing we discover our machine is vibrating too much, we should decrease the jerk and acceleration values.

That was all we need to know about machine movements – please note that we just scratched the surface of motion control, but this is more than enough for most uses.

Before moving forward, I'd like to highlight that, up until now, we've discussed very common parameters found in every **slicing** software, such as Cura or Simplify3D (just to name a couple of valuable alternatives to Fusion 360).

However, since Fusion 360 is CAD software, we can also manage tessellation options – that's what we will discover in the next section.

> **Note**
> Slicing software is responsible for cutting a 3D model and converting it into a set of tiny slices that are then handled by a 3D printer.

Understanding tessellation parameters

First of all, what is **tessellation**? Tessellation is a process where 3D CAD models are converted into 3D mesh models. This conversion is needed for 3D printing in order to calculate toolpaths.

In the following figure, on the left, we have a solid CAD model where the circle profiles are perfectly rounded, and on the right, we converted the model into a mesh via tessellation:

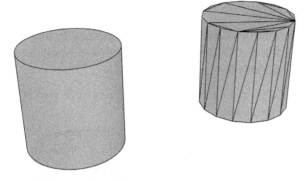

Figure 17.25: CAD geometry versus mesh geometry

The main difference between these two types of 3D geometries is how they are defined; CAD models are based on rigorous math equations and they can describe every shape without losing details. Mesh models, on the other hand, are based on a finite number of vertices and faces connected by edges.

Long story short, tessellation always loses details in the conversion process; that's why it is important to understand how to set it properly. If the tessellation is too rough, we may end up with unforeseen triangles that can ruin the surface finish!

Luckily, there aren't that many parameters we have to truly understand for tessellation:

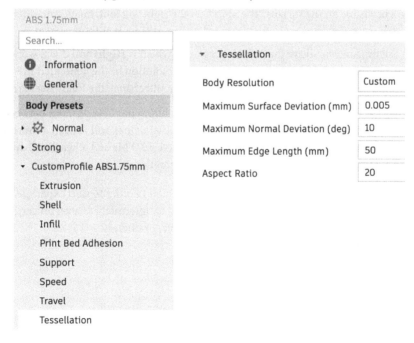

Figure 17.26: Tessellation panel

Let's look at the two most important ones:

- **Body Resolution**: Inside this drop-down panel, we can pick factory presets for tessellation. They are named according to the level of detail – **High** is the most detailed tessellation level, while **Low** is the least detailed.

  In the following figure, we can see differences between two cylinders tessellated with the **Low** preset on the left and the **High** preset on the right:

Figure 17.27: Tessellation differences

  As you can imagine, the object on the left, once printed, won't look like a cylinder; all those edges will be visible on the surface. The object on the right will look much smoother since the edges are so short that they won't be perceived by the naked eye at the first glance.

  Most of the time, we can leave the default preset on (which is **High**) and it will work just fine. However, sometimes we may want to increase the resolution of the generated mesh beyond the maximum default preset. If this is the case, we should change the following parameter.

- **Maximum Surface Deviation (mm)**: This is the maximum possible distance between a CAD surface and its corresponding tessellated mesh. Smaller values will lead to a more accurate mesh. Please note that higher details also mean a heavier 3D file and a heavier G-code file, so check how much data your 3D printer or slicer can handle.

We've now covered all the settings a beginner should understand in order to start experimenting with their own printing settings. We probably even went beyond an intermediate level on certain topics, but I think it was important to mention all these options and parameters.

Summary

That was the end of the chapter! We covered a lot of information, so let's recap what we learned.

First of all, we discovered how to create a custom printing preset by duplicating a default preset. Then, we went through all the most important options to pay attention to when customizing a printing profile.

We discovered how to set the printing temperature for our material, controlling both the extruder and the heated bed, then we found out how to set the layer height, the filament diameter, and an extrusion multiplier to adjust the extrusion width.

After these general settings, we analyzed in detail how to control the shell geometries and the infill geometries, reviewing most of the available settings.

We also looked at advanced adhesion techniques to avoid issues related to thermal shrinkage and temperature differential.

Following this, we went through support material creation and speed and acceleration parameters.

Lastly, we covered the tessellation process and the difference between mesh models and CAD models.

That was quite a valuable list of topics, in my opinion. I hope that all the contents explained up to this point will prove useful for your future applications.

You are almost at the end of the book, but I invite you to move on to the next chapter, where we will test our knowledge with a simple quiz on all the topics covered up to this chapter.

Part 5 – Testing Our Knowledge

This is the last part of the book – after learning so much from previous chapters, it is time to test our knowledge with a quiz covering the entire content of the book.

This part includes the following chapter:

- *Chapter 18, Quiz*

Reading a book where all of the topics are covered and explained can often give you a false sense of having understood everything. However, in reality, there are probably things you may not have properly grasped and need to revisit. A test, in my opinion, is the best way to expose these gaps.

That's what we're going to do now—there are questions on a variety of topics: some are more simple and some are more complex. So, let's see whether we have improved our knowledge of all the subjects covered by this book.

Technical requirements

It is strongly recommended to have read all the chapters!

Questions

Turning (from Chapters 1 to 5)

1. In terms of turning, what is the difference between cutting depth and cutting feed?

    A. The cutting depth is a measure of how much the tool is plunging inside the stock and it is measured in millimeters (mm), while the cutting feed is a measure of how much the tool is advancing at every revolution, and therefore it is measured in mm per revolution (mm/rev).

    B. The cutting depth is how deep the tool is cutting, measuring its position from the rotation axis of the chuck, while the cutting feed is a measure of how fast a tool is plunging radially into the stock.

    C. The depth of cut is a measure of how much the cutting edge is engaged into the stock, while the cutting feed is the tangential velocity measured at the tooltip.

2. What is K_c?

    A. K_c specifies the cutting power needed to cut a certain material, therefore it is measured in kilowatts (kW).

    B. K_c is the specific cutting force, measured in Megapascal (MPa), and is a value representing how easily a certain material can be cut.

    C. K_c is a dimensionless safety factor used in equations that considers power losses due to friction.

3. How does the nose radius of a turning tool affect the surface finish?

    A. A larger nose radius leads to a smoother surface finish.

    B. The surface finish is affected by the feed step and depth of cut only.

    C. A smaller nose radius leads to a better surface finish compared to a larger nose radius.

4. What is the rake surface of a tool?

    A. It is the face formed between the nose edge and the cutting edge.

    B. It is the area where the jaws of the chuck hold the part in place.

    C. It is the surface that the forming chip rubs against.

5. Which rake angle gives the sharpest tool?

    A. A positive rake angle means a sharper tool.

    B. A negative rake angle means a sharper tool.

    C. The rake angle doesn't affect the tool's sharpness.

6. Why do certain lathes have two chucks?

    A. Having two chucks means more power, therefore the cutting speed can be higher for higher productivity.

    B. Generally speaking, one is used to turn parts with smaller radii, while the other is used for bigger components.

    C. Having two chucks allows the part to be moved from one chuck to the other, to machine both sides of the stock without needing to manually flip the stock.

7. Is it possible to flip one axis of the coordinate system without interfering with the other axis?

    A. Yes, it is always possible.

    B. It is possible, but changing the work coordinate system (WCS) requires a new setup.

    C. It is not possible; it would require moving from a right-handed WCS to a left-handed WCS.

8. What is the Safe Z plane?

    A. If properly set, it is a plane where the tool doesn't risk colliding with the stock or the chuck holding the part.

    B. It is a safe area on the machine's working volume where the operator can install a new cutting insert.

    C. It is the position of the chuck jaws.

9. What is defeaturing?

    A. Defeaturing is a modeling technique that aims at simplifying a 3D model (it is often needed for computer-aided manufacturing (CAM) and finite element method (FEM) tasks).

    B. Defeaturing is a way to add CAM operations to a computer-aided design (CAD) model.

    C. Defeaturing is the conversion from CAM operations to the G-code program performed by the postprocessor.

10. What is G-code?

    A. G-code is a programming language that allows users to program and control a CNC machine.

    B. G-code is the naming convention for standardized tools.

    C. G-code is a programming language used to create custom 3D models using the shell command line.

11. What is the ISO-code of a tool?

    A. G-code and ISO-code are the same thing.

    B. It is a standardized set of parameters to use when cutting a part.

    C. It is a naming convention of standardized tools. It is useful to quickly share the most important tool specifications without listing them all.

12. What is CoroPlus?

    A. It is a third-party plugin by Sandvik that grants access to a vast range of cutting tools.

    B. It is an open source library of turning tools.

    C. It is a built-in library of tools that Fusion 360 is based on.

13. What is a potential cause of troubles for facing operations?

    A. Facing operations are risky since the tool is cutting very close to the chuck, therefore collisions may occur if the Safe-Z plane is not set properly.

    B. Facing is a cutting operation where the radial coordinate of the tool can greatly change, therefore the spindle speed must be adjusted accordingly. When cutting close to the center, we must increase the turning speed.

    C. A facing operation requires the tool to cut complex profiles, therefore we must be sure to solve any accessibility issues first.

14. Why should we flag the option to reduce feed when performing the final stock cut?

    A. A reduced feed allows for a better surface finish.

    B. When cutting a part, it gets weaker and weaker the closer the tool arrives to the rotation axis. Therefore, reducing the feed when close to the centerline helps in reducing vibrations and bending.

    C. When cutting a part, chip evacuation is not an issue, therefore we can safely reduce the cutting feed for higher productivity.

15. What are rapid movements?

    A. Rapid movements are a complex set of movements needed to automatically change the tool.

    B. Rapid movements are tool movements performed in the fastest way possible; they are used when not cutting the stock to maximize production.

    C. Rapid movements allow a faster stock cut. This is done by increasing the speed of the machine's motor over safe levels.

16. What is a postprocessor?

    A. It is a CAM software that let us set every cutting operation such as facing, contouring, and so on.

    B. It is part of the CNC machine that helps the user to manually control the machine.

    C. It is a piece of software responsible for translating CAM operations into G-code readable by our machine.

17. How can we prevent a new CAM operation from machining an already machined geometry?

    A. We have to enable the **Clearance** plane to prevent the tool from trespassing into already machined areas.

    B. The option called **Rest Machining** considers previously machined areas and prevents them from being machined again.

    C. There isn't such a risk.

18. What is the most important thing to understand when machining threads?

    A. We must know the target roughness—a coarse surface finish of the thread will likely jam the screw.

    B. We should always know all the thread specifications from standardized tables.

    C. We must carefully pick the threads in the right order—smaller threads will always be machined first.

19. Why is threading a demanding operation on the tool?

    A. Because chip evacuation may become a real problem, especially if the hole is deep and blind.

    B. Because if using constant depth-of-cut passes, the cutting edge is engaged more and more, increasing the tool load. That's why advanced CAM operations allow optimized cutting depths.

    C. Both A and B are right.

20. Why can we drill holes only on the spinning axis in turning?

    A. Because the tools can only be mounted radially.

    B. Because there is not a dedicated spindle for the drill bit—the chuck that holds the part in place is used for drilling, therefore it is possible to drill only on the rotation axis.

    C. Because that would require a custom-made tool.

Milling (from Chapters 6 to 10)

21. What is the difference between table feed and cutting feed?

    A. Table feed is how far the table moves after a single spindle revolution, therefore it is measured in mm/rev. Cutting feed is how fast the tool plunges into the stock, therefore it is measured in **mm per second (mm/sec)**.

    B. There are multiple feed values in milling. Table feed represents how fast the tool is moving forward, measured in **mm per minute (mm/min)**. Cutting feed is a proper feed, representing measures of how far the tool will move after a single tool revolution and is measured in mm/rev.

    C. They are the same thing.

22. What is the difference between up milling and down milling?

    A. The difference is how the tooth engages the stock. Down milling is always preferred over up milling, as the teeth enter where the forming chip is thick. In contrast, with up milling, the teeth enter where the forming chip is thin.

    B. Down milling, also called conventional milling, means that the tool is plunging into the stock axially, moving downward. Up milling is obtained when the tool plunges into the stock, moving upward.

    C. They are the same thing.

23. Which entering angle would you pick for a high-feed facing operation?

    A. A small entering angle such as 15°.

    B. The best entering angle for a facing operation is always 90°.

    C. It depends on the material to be machined.

24. Which is the best entering angle for shoulder milling and face milling?

    A. Shoulder milling requires an entering angle of 15°, while face milling requires an entering angle of 90°—these operations cannot be performed by the same tool!

    B. A tool with a KAPR angle of 45° is always preferred for both face milling and shoulder milling.

    C. Since shoulder milling requires 90° milled faces, we have no choice but to use a 90° KAPR angle.

25. What are the advantages of thread milling against taping?

    A. Thread milling can be performed both with a lathe and a milling CNC center using the same tool.

    B. Thread milling allows a single tool to create multiple threads—it is better for chip evacuation and it can machine a thread almost until the end of a blind hole.

    C. Thread milling can be performed on old machines manually controlling the tool.

26. What is an undercut?

    A. It is part of the stock that has to be cut before moving to the next cutting operation.

    B. It's an inaccessible area for a tool that cannot be machined.

    C. It is another name for the down-milling process.

27. Is it possible to overcome undercut issues?

    A. Yes—there are multiple strategies, from part redesign to a different machining placement.

    B. Yes, using a different tool.

    C. No—the only way is to redesign the part shape.

28. Can an end mill machine a 90° corner?

    A. Yes, but it depends on the orientation of the tool and the part geometry.

    B. Yes, with a multi-axis machine.

    C. No—it will always leave the mill radii on the machined area.

29. In order to manage the left-over mill radius, is reducing the mill diameter an effective strategy?

    A. Reducing the tool diameter may be a solution but sometimes it is not enough to completely solve the problem.

    B. Yes, it will always solve the problem.

    C. Yes, reducing the mill radius will also increase the productivity of the process.

30. What a WCS offset is used for?

    A. A WCS offset is used to tilt a tool at different angles. This is the most common approach to solving undercuts with a multi-axis machine.

    B. A WCS offset allows the machine to reset its coordinates by moving the coordinate system origin to a new location.

    C. It is a machining offset used to ensure a face is entirely machined, in case the stock is bigger than the theoretical dimensions.

31. What is the meaning of G54 in most G-code flavors?

    A. G54 is the default WCS offset.

    B. G54 is the position of the default tool slot for tool change.

    C. G54 is the typical command that enables rapid movements.

32. Why may we want to use a less-than-optimal milling tool for a facing operation?

    A. If the tool is already worn, we may decide to complete its life cycle with a final operation before throwing it into the bin.

    B. If our machined part features both facing operations and shoulder operations, we may prefer to use a single tool for both.

C. Since on most CNC centers we cannot install every tool shape, we may be forced to use a less-than-optimal tool.

33. What is important to check alongside the cutting power required for a certain operation?

A. The total number of cutting edges of the chosen tool.

B. The torque requirements.

C. The cutting insert coating.

34. What is the difference between an adaptive milling command and a non-adaptive milling command?

A. Adaptive commands are capable of autonomous toolpath generation, while non-adaptive commands require human intervention.

B. Adaptive commands create a more advanced toolpath capable of delivering a constant load on the tool, reducing vibrations and resonances.

C. Adaptive commands can automatically change the tool according to the geometry being machined on a single operation; non-adaptive commands can use a single tool per operation instead.

35. What is pecking?

A. It is a strong noise caused by part resonances; it can heavily damage the tool and ruin the surface finish of the part.

B. It is a chip management strategy for deep holes.

C. It is a custom drill bit featuring coolant channels for deep drilling.

36. Why can we not be sure whether a toolpath is cutting in up or down milling just by looking at it?

A. Because we should also know the cutting-edge placement and the spinning direction of the tool.

B. Because we don't know the entry point and the exit point of the toolpath.

C. Because the toolpath doesn't show the tool.

37. Why can a ball nose mill cause issues on a three-axis CNC?

A. Because it can only be used on multi-axis machines.

B. A Cartesian milling machine cannot tilt the tool; therefore, when milling flat horizontal faces, the cutting speed of the tool on the tip would be 0.

C. Because it requires feeds and powers that only a multi-axis machine can sustain for a prolonged time.

38. What should we check when milling a hole with a flute end mill?

    A. We must check that both the maximum helix angle and the maximum cutting depth requirements are respected.

    B. We should check for collisions against the thread's tooth.

    C. A flute mill is a special tool used for threading; therefore, we must understand the thread pitch and its diameters.

39. What is roughness, R_z ?

    A. It is a scientific method to evaluate the roughness of a part based on the amount of reflected light. The shinier the surface, the more light is reflected. Measuring the reflection intensity gives the roughness of the surface.

    B. It is the roughness measured along the z axis of the active WCS offset.

    C. It is a rather common way of measuring roughness by comparing the highest peaks and the deepest valleys of a surface.

Laser cutting (from Chapters 11 to 13)

40. Laser cutting beams are:

    A. Visible and deadly.

    B. Invisible and deadly.

    C. Nice to play with.

41. How does a laser cut?

    A. A laser beam cuts a metal sheet when focused on a tiny spot on the surface. When melted (or vaporized), the metal can be evacuated from the bottom via a strong jet of auxiliary gases.

    B. A laser beam cuts a metal sheet by heating it beyond the melting point; the melted metal is then evacuated from the bottom thanks to gravity.

    C. Thanks to the reflected light, it is possible to cut a part.

42. Which technology can cut a metal sheet 2 mm thick the fastest?

    A. Milling.

    B. Laser cutting.

    C. Plasma cutting.

43. What is calamine?

    A. It is a dark layer of oxide created on top of metal sheets, caused by the sparks generated during the cutting process.

    B. It is the typical shape profile created on the bottom of thick plates cut by lasers.

    C. It is an unwanted layer of black oxide formed on the cut surface, caused by the reaction between oxygen and steel while cutting.

44. What is nesting?

    A. Nesting is a process that aims at reducing wasted material by optimizing part placements.

    B. Nesting is the possibility of tilting the laser cutting head to create complex 3D profiles on thick metal sheets.

    C. Nesting is an automatic process to manage the metal sheets' warehouse. Using nesting, we can reduce the amount of metal scrap generated.

45. What is sheet stacking?

    A. Sheet stacking means cutting multiple sheets at once stacked together. It is a common technique for waterjets; however, it is impossible for lasers since the extreme heat would weld the sheets together.

    B. Sheet stacking means cutting multiple sheets at once stacked together. It is a common technique for lasers and can vastly increase productivity.

    C. Sheet stacking means cutting multiple sheets at once stacked together. It is a common technique for waterjets. Thanks to stacking, it is possible to cut different profiles on every stacked sheet.

46. Is there one way only for nesting optimization?

    A. Yes—there is only one way of creating a nesting. It is like playing *Tetris*—the more parts you can fit together, the better!

    B. Yes—the best nesting type is obtained when optimizing for large production batches.

    C. No—it is possible to optimize nesting both for kit optimization and batch optimization (both have their pros and cons).

47. What are tabs on a nested part used for?"

    A. Tabs are used to perfectly align the metal sheets for stacking.

    B. Tabs are interrupted cuts; they are used to let the part cool down and avoid distortions in the metal lattice.

    C. Tabs are micro joints between the part and the metal sheet; they are needed to prevent parts from flipping and exposing the laser head to dangerous impacts.

48. How would you describe the width of a laser cut?

    A. Zero—laser cutting generates no chip, in fact.

    B. Very small, but has to be considered to get proper tolerances on the parts.

    C. Larger on thinner metal sheets.

Additive manufacturing (from Chapters 14 to 17)

49. Which is the best additive manufacturing technology?

    A. **Stereolithography (SLA)** is the best since it can print multiple materials together.

    B. **Fused deposition modeling (FDM)** is the best since it has the best surface quality and doesn't need support structures.

    C. It depends on the needs of the user; there is no overall winner.

50. Is SLA an eco-friendly printing process?

    A. Not really—stereolithography relies on chemical resins that often need to be washed with alcohol, resulting in special waste that is hard to recycle.

    B. Yes—most PLA filaments are polymerized by corn; therefore, they are completely recyclable.

    C. No—nylon powder is noxious when exposed to the chemical bath to complete its polymerization.

51. Why does selective laser sintering (SLS) have the highest productivity among other 3D printing technologies?

    A. Because it uses support structures capable of holding multiple parts together.

    B. Because it can print bigger components faster than other technologies.

    C. Since it doesn't require support structures, it is possible to stack components one on top of the other to entirely fill the build volume.

52. Which is the only additive manufacturing process capable of printing multiple materials at once?

    A. FDM—it can use multiple extruders in parallel.

    B. SLS—most of the resins are carbon fiber reinforced.

    C. SLA—it can use multiple extruders in parallel.

53. Is it possible to print parts with undercuts using the FDM printing technique?

    A. No—much like milling, undercuts have to be removed before proceeding with the printing process.

    B. Yes—however, extreme undercuts may require additional support structures.

    C. Yes—3D printers don't experiment with undercuts since they print the object layer by layer.

54. Does layer orientation affect the mechanical properties of an FDM-printed part?

    A. No—the material is polymerized between the layers, therefore the resulting object is isotropic.

    B. Yes, since it may require additional support structures to reinforce the part strength.

    C. Yes—the printed part may feature a severe anisotropy perpendicular to the layer orientation.

55. What does material anisotropy mean?

    A. We have anisotropies when a material features different mechanical properties along different directions.

    B. It's a different cooling behavior caused by the shape of a part. This different cooling speed creates residual tensions inside the structure that weaken the part in certain directions.

    C. It's a term used to express the strength of composite materials like carbon fibers.

56. Is a printed part affected by thermal expansion?

    A. No—printing processes happen at room temperature; therefore, even if there is a tiny thermal expansion, we can safely ignore it.

    B. Yes—the printing material is extruded at high temperatures; therefore, when cooling down, the part will shrink, which may cause bad dimension tolerances and distortions.

    C. Yes, but only if printing two different materials. Different materials have different expansion coefficients, in fact.

57. Which additive technology doesn't need support structures?

    A. SLS.

    B. FDM.

    C. SLA.

58. What is generative design?

    A. It is an advanced design process that creates a part to optimize its production costs.

    B. It is a new design approach where a shape is created by a set of programmed instructions.

    C. It is a design approach where a part is generated by software that optimizes the shape according to the boundary conditions.

59. Does additive manufacturing have the potential for mass production?

    A. No, and it probably never will. It lacks scalability.

    B. No, additive manufacturing is too expensive.

    C. Yes, thanks to its great flexibility and scalability.

60. Which additive manufacturing process can print multiple materials at once?

    A. FDM.

    B. SLA.

    C. SLS.

61. Which of the following is a printing technique to improve bed adhesion for an FDM printer?

    A. Pecking.

    B. Skirt.

    C. Brim.

62. What is a gyroid?

    A. A common infill pattern for FDM printers.

    B. A type of support structure.

    C. A flexible filament that needs higher extrusion flows to be printed.

63. What is jerk?

    A. It is a value that describes how fast the acceleration changes.

    B. It is a value that describes how fast the speed changes.

    C. It is a value that describes how fast the extruder moves.

64. What is tessellation?

    A. It is a conversion from CAD data to G-code handled by the postprocessor.

    B. It is a conversion from CAD data to mesh data.

    C. It is a conversion from mesh data to CAD data.

65. What is the typical extrusion temperature for ABS filaments?

    A. 85°C.

    B. 300°C.

    C. It depends on the material specs, however it is around 200°C.

Answers

Turning

1A	2B	3A	4C	5A	6C	7C	8A	9A	10A
11C	12A	13B	14B	15B	16C	17B	18B	19C	20B

Milling

21B	22A	23A	24C	25B	26B	27A	28A	29A	30B
31A	32B	33B	34B	35B	36A	37B	38A	39C	

Laser cutting

40B	41A	42B	43C	44A	45A	46C	47C	48B

Additive manufacturing

49C	50A	50A	52A	53B	54C	55A	56B	57A
58C	59A	60A	61C	62A	63A	64B	65C	

Summary

Congratulations—with that final quiz, we have found ourselves at the end of the book!

We covered many topics, starting from turning and milling to laser cutting and additive manufacturing. Not only could we explore most of the CAM environment of Fusion 360 in detail, but I'm also glad that we could dive into more complex technical aspects that will for sure improve the design approach of any beginner.

Thanks for having joined me on this long journey—I hope it was worth it. I wish you the best!

Index

Symbols

2D Adaptive Clearing command 233
Geometry tab 234-236
Passes tab 236, 237
Tool tab 233

3D-printed component, limitations 355
anisotropies 361
bed adhesion 360
overhang geometries, printing 356

3D printing
cons 346
facing overhangs 357, 358
pros 343

3D printing, cons
limited part dimensions 348
limited range of materials 348
scalability 346-348

3D printing, pros
complex shapes 343
composites 345
just in time supplying 345
rapid prototyping 344

A

acceleration 293, 407
Accessibility Analysis command 178, 179
actors, turning
chip 4
chuck 4
cutting tool 4
machined stock 4

Adaptive Clearing
Geometry tab 256
Heights tab 256-258
Linking tab 260-262
Passes tab 258-260
Tool tab 255

additive manufacturing 342
anisotropies 309
of printed part 361

Arrange command 302-304
using 306, 307

Automatic Orientation 374
Parameters tab 375
Ranking tab 375-379

auxiliary cutting edge 19
auxiliary flank 21

B

backside milling 184
batch volume 300, 301
bed adhesion 360
bending 27, 28
Bore
    Geometry tab 264
    Heights pane 264, 265
    Passes tab 265-268
    Tool tab 263

C

calamine 295
CAM project
    interface 32, 33
Cartesian coordinates 5
Cartesian machines 160, 161
chuck
    importing 35-38
Chuck subpanel
    Chuck reference 53
    Offset value 53, 54
climb milling 166
CO2 lasers 291
components 304, 305
composite 345
computer numerical control
        (CNC) machine 32
conventional milling 166
CoroPlus 77
    new tools, importing with 78-83
    used, for calculating cutting
        parameters 87-91
    used, for finding tool for face milling
        and shoulder milling 216-222
counterboring 143

Cutter tab 328
    kerf width 329
    nozzle diameter 328
Cutting data tab 329
    Assist gas 330
    Cut height 330
    Cut Power 330
    Cutting feedrate 330
    Pierce height 330
    Pierce power 330
    Pierce time 330
    Pressure 330
cutting depth 11, 163
    axial depth of cut 163, 164
    radial depth of cut 163, 164
cutting edge 19
cutting feed 11-13
cutting operation
    implementing 330
cutting parameters 162
    calculating, with CoroPlus 87-91
    cutting depth 163
    cutting power 168, 169
    cutting speed 162, 163
    cutting torque 168, 169
    entering, into Fusion 360 91, 92
    feed step 164
    spindle speed 162, 163
cutting power 13, 14, 168, 169
cutting speed 10, 11, 162, 163
cutting tool
    creating 326, 327
    Cutter tab 328
    Cutting data tab 329
    Geometry tab 332, 333
    Height tab 335
    simulation result 336, 337
    Tool tab 331, 332

cutting torque 168, 169
cylindrical coordinates 5, 6

D

defeaturing 51
discretization 99
down milling 166
draft analysis 179
Drilling 140, 238
  Cycle tab 142, 242-244
  Geometry tab 141, 142, 240
  Heights tab 240-242
  Tool tab 140, 141, 239

E

electromagnetic (EM) wave 290
electromagnetic spectrum 290
entering angle 26-28
example part 298, 299
  model, presenting 365, 366
exothermic reaction 295
external machining 7
Extruder 1 options 394, 395
extruder parameters
  reviewing 392, 393
extrusion options 393, 394

F

Face command
  Geometry tab 226
  Passes tab 226-229
  Tool tab 224, 225
face milling 170, 171, 216, 250, 251
  CoroPlus, used for finding best tool 216-223
  implementing, with Fusion 360 224

facing operation
  CoroPlus, for calculating cutting
    parameters 87-91
  cutting parameters, entering
    into Fusion 360 91-94
  setting up 86, 87
facing operations 16
facing overhangs
  in 3D printing 357, 358
feed per revolution 166-168
feed per tooth 165, 166
feed step 164
  feed per revolution 166-168
  feed per tooth 165, 166
  table feed 168
fiber lasers 291
first layer placement
  selecting 362-364
Fused Deposition Modeling (FDM) 349
  advantages 349
  disadvantages 350
Fusion 360
  example projects, finding 33-35
  using, for laser cutting 321, 322

G

G-code 59, 60
  checking 111, 112
  reference link 116
general parameters
  reviewing 390-392
generative design 343
Geometry tab 332, 333
  options 333

H

High-Speed Steel (HSS) 221
hole
　milling 262

I

infill parameters 398-400
infrared (IR) 291
internal machining 7

J

jaws 36
jerk 408
just in time supplying 345

K

kerf width 329
kilowatts (kW) 13

L

Lantek
　URL 292
laser cutting 293
　advantages 293, 294
　drawbacks 294-296
　Fusion 360, using 321, 322
　Post Process tab 325
　setup, creating 322
　Setup tab 323, 324
　Stock tab 324
laser-cutting machine
　working 291-293
lasers 289-291

lathe 4
longitudinal machining 15
loop options
　all loops 333
　inner loops 333
　outer loops 333

M

machining issues, solving 191
　counterbore holes 195, 196
　fillets 197
　missing tool radii 194
　undercut face 191
machining strategies 15
　facing operations 16
　longitudinal operations 15
　plunging operations 16, 17
　profiling operations 17, 18
main flank 21
manufacturing model 317
meters over seconds squared (m/s2) 293
meters per minute (m/min) 10, 134, 162
microwaves 290
millimeters (mm) 10
millimeters per revolution (mm/rev) 11, 166
milling 160
　cartesian machines 160, 161
　multi-axis machines 161, 162
　working 160
milling operations 170
　face milling 170, 171
　other types 175, 176
　profile milling 175
　shoulder milling 172
　simulation, running 246, 247
　slot milling 173, 174

mill radius

cutting direction, changing 189, 190

managing 185

part geometry, tweaking 188, 189

reducing 186, 187

Minimize Build Height command 372-374

model

orienting, onto build platform 370

Model subpanel

Model 50

Spun Profile 51, 52

Morphed Spiral

Geometry tab 271, 272

Heights panel 272

Passes tab 273-275

Tool tab 270, 271

using 269, 270

multi-axis machine

used, for undercut analysis 181

multi-axis machines 161, 162

multiple setup

versus single setup 204

multi-start threads

reference link 137

N

nesting 294, 299

Arrange command 302-304

batch volume 300, 301

Global Parameters panel 315, 316

manual placement 302

Nesting and Fabrication extension 307

Output panel 316-319

Packaging tab 314, 315

Shape tab 313

sheet format 299, 300

Study tab 312

with Fusion 360 302

Nesting and Fabrication extension 307

material and sheet format 307-310

part list 310-312

Newton meter (Nm) 169

nose angle 24, 25

nose edge 19, 20

nozzle diameter 328

O

overhangs 375

P

part 200

part catcher 152

part fixture

selecting 202-204

part nesting

creating 312

part orientation

selecting 362-364

part placements

selecting 201, 202

piercing 330

Place parts on platform command 371, 372

plugin 77

plunging operations 16, 17

polymerization 351

Post Process command 112

operations 114-116

settings 112-114

postprocessing 59

post-processor 59

using 384, 385

Post Process tab 59, 209, 210, 212, 213, 325
  Machine WCS 61
  Program Comment 61
  Program Name/Number 61
print bed adhesion 400-404
printing preset
  creating 388-390
printing setup
  creating 366-370
prints
  improving, with support structures 358, 359
profile milling 175
profiling operations 17, 18

Q

quenching 295

R

radio waves 290
rake angle 22
rake surface 21
rapid prototyping 344
relief angle 23
revolutions per minute (RPM) 10, 162
rotation per minute (rpm) 134
roughing operation
  implementing, with adaptive
    clearing 251-254
Round-tool KAPR 28, 29

S

Safe Z
  Offset 49, 50
  Safe Z Reference 47, 48

sample tool
  obtaining 66, 67
Selective Laser Sintering (SLS) 352, 353
  advantages 353
  disadvantages 353
setup 39
Setup subpanel
  Operation type 41
  Spindle 42
Setup tab 40, 41, 207-211, 323, 324
  Chuck subpanel 52
  Machine 41
  Model subpanel 50
  Safe Z 47
  Setup subpanel 41
  Work Coordinate System (WCS) 43-47
sheet format 299, 300
sheet stacking 313
shell parameters 395-398
shoulder milling 172, 229
  centerline placement 172, 173
  CoroPlus, used for finding best tool 216-223
  cutting parameters, evaluating 229-232
  in Fusion 360 233
  milling diameter, selecting 172
Simulate interface
  current operation 104
  progress bar 104
  SIMULATE panel 104
  simulation controls 105
  speed cursor 105
  stock 104
  tool 104
single setup
  versus multiple setup 204
slot milling 173, 174
  tasks 174

speed parameters 405, 406
  acceleration 407, 408
  jerk 408
  speed 407
spindle 42
spindle orientation 189
spindle speed 162, 163
stereolithography (SLA) 350-352
  advantages 351
  disadvantages 351, 352
Stock tab 54, 55, 208, 209, 211, 324
  Fixed size box 56
  Fixed size cylinder 58
  Fixed size tube 58
  From preceding setup 59
  From solid 58
  Relative size box 56, 57
  Relative size cylinder 58
  Relative size tube 58
support material parameters 404, 405
support structures
  generating 379-381
  used, for improving prints 358, 359

T

table feed 168
tabs 334
  options 334, 335
tapping 143, 244
  drilling's Cycle tab, used 246
  drilling's Heights tab, used 245
  drilling's Tool tab, used 245
tessellation 408
  parameters 408-410
thermal expansion 401
thermal shrinkage 401

third-party tool library
  importing 74-77
thread geometry 276-279
threading tool
  selecting 279
thread milling 275
  Geometry tab 282
  Heights tab 283
  implementing, in Fusion 360 280
  Passes tab 284, 285
  Tool tab 280, 281
tool
  cloning 67, 68
  creating 65
tool angles 21
  entering angle 26-28
  nose angle 24, 25
  rake angle 22
  relief angle 23
tool edges 18, 19
  auxiliary cutting edge 19
  cutting edge 19
  nose edge 19, 20
tool flanks
  auxiliary flank 21
  main flank 21
tool geometry 18
  usages 25, 26
tool library 64, 65
  Cutting data tab 71, 72
  General tab 68, 69
  Holder tab 70
  Insert tab 69
  Post Processor tab 73
  Setup tab 71
toolpath
  simulating 382-384

tool simulation 102
    discovering 102-111
    simulated result 105
tool surfaces 20, 21
    rake surface 21
    tool flanks 21
Tool tab 331, 332
turning 4
Turning Face 93
    Geometry tab 94, 95
    Passes tab 99-101
    Radii tab 95-98
    Tool tab 93, 94
Turning Groove 144, 145
    Geometry tab 147, 148
    Passes tab 148, 149
    Simulation 150
    Tool tab 145-147
turning operations 6
    external machining 7
    internal machining 7
turning operations, parameters 8
    cutting depth 11
    cutting feed 11-13
    cutting power 13, 14
    cutting speed 10, 11
    turning speed 9
Turning Part 150, 151
    Geometry tab 152-154
    Passes tab 154, 155
    Radii tab 154
    Simulation 156
    Tool tab 151, 152
Turning Profile (Finishing) 126, 127
    Geometry tab 128
    Passes tab 128, 129
    Simulation 130
    Tool tab 127, 128

Turning Profile (Roughing) 118-120
    Geometry tab 123
    Passes tab 124, 125
    Simulation 125, 126
    Tool tab 120-122
turning speed 9
Turning Threads 130-132
    Geometry tab 134, 135
    Passes tab 136-138
    Radii tab 136
    Simulation 138, 139
    Tool tab 133, 134

U

undercut analysis
    backside milling 184
    custom tools, creating 182-184
    multi-axis machine, using 181
    part orientation, changing 180, 181
undercut face, solving
    with custom tool 193, 194
    with deep milling 193
    with inclined tool 191, 192
undercuts 356
    handling 178, 179
up milling 166

W

Work Coordinate System
    (WCS) 33, 43-47, 205
    offsets, selecting 204-206

www.packtpub.com

Subscribe to our online digital library for full access to over 7,000 books and videos, as well as industry leading tools to help you plan your personal development and advance your career. For more information, please visit our website.

Why subscribe?

- Spend less time learning and more time coding with practical eBooks and Videos from over 4,000 industry professionals

- Improve your learning with Skill Plans built especially for you

- Get a free eBook or video every month

- Fully searchable for easy access to vital information

- Copy and paste, print, and bookmark content

Did you know that Packt offers eBook versions of every book published, with PDF and ePub files available? You can upgrade to the eBook version at packtpub.com and as a print book customer, you are entitled to a discount on the eBook copy. Get in touch with us at customercare@packtpub.com for more details.

At www.packtpub.com, you can also read a collection of free technical articles, sign up for a range of free newsletters, and receive exclusive discounts and offers on Packt books and eBooks.

Other Books You May Enjoy

If you enjoyed this book, you may be interested in these other books by Packt:

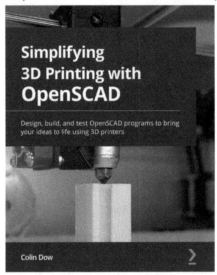

Simplifying 3D Printing with OpenSCAD

Colin Dow

ISBN: 978-1-80181-317-4

- Gain a solid understanding of 3D printers and 3D design requirements to start creating your own objects
- Prepare a 3D printer for a job starting from leveling the print bed and loading the filament
- Discover various OpenSCAD commands and use them to create shapes
- Understand how OpenSCAD compares to other CAD programs
- Get to grips with combining text and a cube to create an object
- Explore the common libraries in OpenSCAD

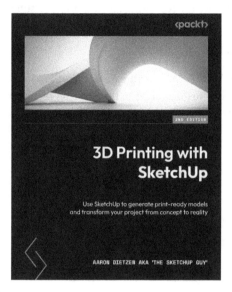

3D Printing with SketchUp

Aaron Dietzen

ISBN: 978-1-80323-735-0

- Understand SketchUp's role in the 3D printing workflow
- Generate print-ready geometry using SketchUp
- Import existing files for editing in SketchUp
- Verify whether a model is ready to be printed or not
- Model from a reference object and use native editing tools
- Explore the options available for adding onto SketchUp for the purpose of 3D printing (extensions)
- Understand the steps to export a file from SketchUp

Packt is searching for authors like you

If you're interested in becoming an author for Packt, please visit authors.packtpub.com and apply today. We have worked with thousands of developers and tech professionals, just like you, to help them share their insight with the global tech community. You can make a general application, apply for a specific hot topic that we are recruiting an author for, or submit your own idea.

Share Your Thoughts

Now you've finished *Making Your CAM Journey Easier with Fusion 360*, we'd love to hear your thoughts! Scan the QR code below to go straight to the Amazon review page for this book and share your feedback or leave a review on the site that you purchased it from.

https://packt.link/r/1-804-61257-X

Your review is important to us and the tech community and will help us make sure we're delivering excellent quality content.

Download a free PDF copy of this book

Thanks for purchasing this book!

Do you like to read on the go but are unable to carry your print books everywhere?

Is your eBook purchase not compatible with the device of your choice?

Don't worry, now with every Packt book you get a DRM-free PDF version of that book at no cost.

Read anywhere, any place, on any device. Search, copy, and paste code from your favorite technical books directly into your application.

The perks don't stop there, you can get exclusive access to discounts, newsletters, and great free content in your inbox daily

Follow these simple steps to get the benefits:

1. Scan the QR code or visit the link below

https://packt.link/free-ebook/9781804612576

2. Submit your proof of purchase
3. That's it! We'll send your free PDF and other benefits to your email directly

www.ingramcontent.com/pod-product-compliance
Lightning Source LLC
Chambersburg PA
CBHW081457050326
40690CB00015B/2831